Quantum Physics for Beginners

Easy Guide to Learn the Basic Concepts and the Secrets of the Universe with the Most Important Theories and Intuitive Examples

Edwin Futrell

© Copyright 2021 by (Edwin Futrell)- All rights reserved.

This document is geared towards providing exact and reliable information about the topic and issue covered. The publication is sold with the idea that the publisher is not required to render accounting, officially permitted, or otherwise qualified services. If advice is necessary, legal, or professional, a practiced individual in the profession should be ordered.

- From a Declaration of Principles which was accepted and approved equally by a Committee of the American Bar Association and a Committee of Publishers and Associations.

In no way is it legal to reproduce, duplicate, or transmit any part of this document in either electronic means or printed format. Recording of this publication is strictly prohibited, and any storage of this document is not allowed unless with written permission from the publisher. All rights reserved.

The information provided herein is stated to be truthful and consistent, in that any liability, in terms of inattention or otherwise, by any usage or abuse of any policies, processes, or directions contained within is the solitary and utter responsibility of the recipient reader. Under no circumstances will any legal responsibility or blame be held

against the publisher for reparation, damages, or monetary loss due to the information herein, either directly or indirectly.

Respective authors own all copyrights not held by the publisher.

The information herein is offered for informational purposes solely and is universal as so. The presentation of the information is without a contract or any type of guarantee assurance.

The trademarks used are without any consent, and the publication of the trademark is without permission or backing by the trademark owner. All trademarks and brands within this book are for clarifying purposes only and are owned by the owners themselves, not affiliated with this document.

Table of Contents

INTRODUCTION 7

CHAPTER 1: INTRODUCTION TO THE QUANTUM MECHANICS 13
   1.1 EXPLAIN THE DIFFERENCE BETWEEN QUANTUM PHYSICS AND CLASSICAL PHYSICS? 16

CHAPTER 2: FATHER OF THE QUANTUM ENTANGLEMENT 31
   2.1 THE PHOTON'S WAVE-PARTICLE DUALITY 33
   2.2 THE LIGHT IS A PARTICLE? LIGHT QUANTUM SUPPOSITION OF EINSTEIN - 33
   2.3 THE DUALITY OF PHOTONS 35
   2.4 THE NATURE OF THE PHOTON 37
   2.5 "PHOTON ON TRIAL" 38
   2.6 "INVITATION TO THE PHOTON." 40

CHAPTER 3: YOUNG'S DOUBLE SLIT EXPERIMENT 41
   3.1 TAKE-HOME EXPERIMENT: USING FINGERS AS SLITS 46
   3.2 EXAMPLE 1. FINDING A WAVELENGTH FROM AN INTERFERENCE PATTERN 49
   3.3 EXAMPLE 2. EVALUATING HIGH RANKING ORDER POSSIBLE 50
   3.4 SECTION SUMMARY 51
   3.5 CONCEPTUAL QUESTIONS 52

CHAPTER 4: WHAT IS HEISENBERG'S UNCERTAINTY PRINCIPLE? 54

4.1 WHY IS CALCULATING BOTH MOMENTUM & POSITION CONCURRENTLY IMPOSSIBLE? 54
4.2 HEISENBERG UNCERTAINTY PRINCIPLE FORMULA AND APPLICATION 56
4.3 EXPLAINING HEISENBERG UNCERTAINTY PRINCIPLE WITH AN EXAMPLE 58
4.4 HEISENBERG'S Γ-RAY MICROSCOPE 59
4.5 DOES UNCERTAINTY PRINCIPLE OF HEISENBERG NOTICEABLE IN ALL THE MATTER WAVES? 61
4.6 HEISENBERG UNCERTAINTY PRINCIPLE EQUATIONS 61
4.7 EXPLAINED NUMERICAL DIFFICULTIES ON UNCERTAINTY PRINCIPLE OF HEISENBERG 62
4.8 SCHRODINGER'S CAT 66
4.9 THIS TWIST CAT PARADOX OF THE SCHRÖDINGER HAS MAIN INSINUATIONS FOR THE QUANTUM THEORY . 69
4.10 THE MATTER OF THE TASTE 72
4.11 A METHOD TO OBSERVE WIGNER'S FRIEND 76
4.12 RECONSIDERING REALITY 82

CHAPTER 5: EINSTEIN'S THEORY OF RELATIVITY 88

5.1 COVARIANCE, REFERENCE FRAMES & GALILEAN RELATIVITY 88
5.2 THE LORENTZ TRANSFORM & THEORY OF RELATIVITY (EINSTEIN) 96
5.3 TIME DILATION 98
5.4 DEMO: MUON DECAY 101
5.5 LENGTH CONTRACTION 103
5.6 LORENTZ TRANSFORM 104
5.7 DEMO: TRADITIONAL LIMIT 109
5.8 SPACETIME 110

CHAPTER 6: SIMPLE MADE ENTANGLEMENT 118

6.1 QUANTIZED 119

CHAPTER 7: EPR PARADOX IN PHYSICS 133

7.1 HOW EPR PARADOX DESIGNATES THE QUANTUM ENTANGLEMENT 133
7.2 THE PARADOX'S STARTING 133
7.3 THE PARADOX'S SENSE 135
7.4 THEORY OF HIDDEN-VARIABLES 135
7.5 UNCERTAINTY IN THE QUANTUM MECHANICS 136
7.6 BELL'S THEOREM 137

CHAPTER 8: QUANTUM THEORY: EVOLUTION OF QUANTUM THEORY 138

8.1 FIVE PRACTICAL USAGES FOR "SPOOKY" THE QUANTUM MECHANICS 140
8.2 SCIENTISTS CAPTURE SCHRÖDINGER'S CAT AT THE CAMERA 141
8.3 SEVEN SIMPLE METHODS YOU'LL RECOGNIZE EINSTEIN WAS CORRECT (FOR NOW) 141
8.4 MERCURY'S ORBIT 142
8.5 BENDING LIGHT 143
8.6 BLACK HOLES 145
8.7 SHOOTING MOON 146
8.8 DRAGGING SPACE 147
8.9 SPACE-TIME RIPPLES 148
8.10 GPS 149
8.11 PHYSICS CAN BRAND THE INVISIBLE CAT, VISIBLE 150
8.12 ULTRA-PRECISE CLOCKS 151
8.13 UNCRACKABLE CODES 152
8.14 SUPER- COMPUTERS 153
8.15 IMPROVED MICROSCOPES 154

8.16 BIOLOGICAL COMPASSES 155
CHAPTER 9: WHERE, IN THE FUTURE, IS THE QUANTUM COMPUTING? 157
9.1 WHAT IS NEXT IN THE COMPUTING? EXPLORING POSSIBILITIES 157
9.2 QUANTUM WAY 158
9.3 SUPERCONDUCTING CIRCUITS 160
9.4 WHAT CAN YOU DO WITH A QUANTUM COMPUTER? 160
9.5 USING LIGHT VERSUS ELECTRONS FOR PROCESSORS 161
CONCLUSION 163

Introduction

Theory of the Quantum field is seemingly the most sophisticated and excellent physical theory at any point developed, with perspectives more rigidly tried and validated to more noteworthy exactness than some other theory in physical science. Shockingly, the subject has achieved notoriety for being difficult, with precluding looking Math and an unconventional diagrammatic language portrayed in various unforgiving, heavy reading material pointed solidly at inspiring experts. Though, a quantum field theory is excessively significant, excessively delightful, and too captivating to be limited to the experts possibly. This book on the quantum field is intended to appear as something different.

It is composed of test physicists and intends to give the intrigued beginner a scaffold from undergrad physical science to the quantum field theory. The envisioned reader is a skilled beginner, having a curious and versatile brain, hoping to be told some engaging and mentally stimulating story, yet who won't feel belittled if a couple of numerical comforts are illuminated. Utilizing various worked models, graphs, and cautious genuinely spurred clarifications, this book most probably smoothed the way concerning understanding drastically extraordinary and progressive perspective on the actual world that the theory quantum field gives, and which all physicists ought to have the chance to encounter.

Giving a complete prologue to the theory of the quantum field, this text material covers the improvement of molecule physical science from its establishments to the disclosure of Higgs boson. It is a blend of clear actual clarifications, with direct associations with trial information and numerical meticulousness that make the subject open to understudies with a wide assortment of foundations and interests. Expecting just an undergrad-level comprehension of quantum mechanics, this book consistently builds up the Standard Model and cutting-edge estimation procedures.

It incorporates numerous derivations of numerous significant results, with present-day techniques, for example, successful field theory & renormalization group assuming a bulging part. Various worked models and end-of-section issues emperor students to regain exemplary outcomes and dominate the theory of the quantum field as today is utilized. Given a course instructed by the writer over numerous years, this book is perfect for starting the complex theory of quantum field succession or for readers who are independent.

A far-reaching introduction to scientific standards of a mind-perplexing subject where importance and understanding never stop to perplex and astound. A-Z manage, which is neither too progressed nor distorted and which is finished with figures and charts that show the more profound significance of the ideas you are probably not going to discover somewhere else. The strangeness and oddities of quantum physical science are clarified at a basic level, from the main standards to current complex tests.

This is for the non-physicist self-teacher who is searching for general information about quantum physical science, as it outfits the most thorough record that work can give and which just occasionally, in scarcely any extraordinary sections, resorts to a numerical level that goes no farther than that of secondary school. It will save you a huge load of time in looking somewhere else, attempting to bits together an assortment of data. Rather than being 'quantum physical science for dummies,' It's a more profound record that sums up the analyses as You'll talk about the philosophical contentions and reasonable establishments. It is a guide for each one of the individuals who have consistently been pulled into the interesting quantum reality and needed to comprehend its standards, regardless of whether they are not physicists, but rather have discovered just either progressed college-level coursebooks loaded up with complex arithmetic or, then again, mainstream science messages that attempted to associate with the reader at the expense of unrefined distortion.

This book makes up for a shortfall: that of a non-scholarly hoover genuine introduction to the reasonable foundations of quantum material science without dealing with readers like imbeciles. We have attempted to traverse a 'curve of information' without yielding to the allurement of taking an unreasonably uneven record of the subject. With that in mind, we have avoided zeroing in a lot on his inclinations – something that in any case would have ruined the goal of making this an overall presentation. It is all things being equal, above all else, a push to give the reader the broadest conceivable foundation on all the nuts and bolts that everybody keen on quantum physical science ought to have.

What is this? Your thing called quantum material science? What is its effect on your comprehension of the world? What is 'reality' as indicated by quantum material science? This book tends to these and numerous different inquiries through a bit-by-bit venture into this exceptionally abnormal world. The focal secret of the twice slit trial and the wave-molecule duality, the fluffy universe of Heisenberg's vulnerability guideline, the bizarre Schrödinger's feline mystery, the 'creepy activity a good way off' of quantum trap, the EPR Catch 22 and significantly more is clarified, without dismissing such principal givers as Planck, Einstein, Bohr, Feynman, and other people who battled themselves to concoct the strange quantum domain. This manual additionally investigates the trials led in ongoing many years, for example, the amazing "what direction" and "quantum-deletion" tests

Furthermore, on the grounds that universities, schools, and colleges encourage quantum physical science utilizing a dry, for the most part, a specialized methodology which outfits just shallow knowledge into its establishments, this book is prescribed to each one of those understudies and physicists who might want to look past the simply formal viewpoint and dive further into the importance and substance of quantum mechanics.

First*, thanks for trusting me and deciding to choosing this book! I'd really love to hear your thought, so please, leave me a short review if you enjoy it.*

Thanks for spending your time

I have a GIFT FOR YOU at the end of the book to see all the images in color

Chapter 1: Introduction To The Quantum Mechanics

Quantum mechanics is an actual science managing the conduct of energy and matter on the size of subatomic and atomic waves/particles. It additionally frames the reason for the contemporary comprehension of how enormous objects, for example, stars & galaxies, & cosmological occasions, the Big Bang, can be dissected and clarified. Quantum mechanics is establishing a few related orders, including consolidated matter physics, nanotechnology, quantum science, electronics, particle physics, and structural biology. The expression "quantum mechanics" was 1st authored by Max Born in 1924. The acknowledgment by the overall physical science local area of quantum mechanics is because of its exact forecast of frameworks' physical behavior, including frameworks where Newtonian mechanics fizzles. Indeed, even broad relativity is restricted - in manners, quantum mechanics isn't - for depicting frameworks at atomic scale or even smaller, at low or exceptionally higher energies, or the most minimal temperatures. The quantum mechanical hypothesis has been demonstrated to be exceptionally effective and pragmatic during experimentation and applied science.

The establishments of quantum mechanics date from the mid-1800s; you've got the genuine beginnings of QM date from crafted by Max Planck in 1900. Albert Einstein & Niels Bohr soon yielded significant influences on what is currently known as "old quantum theory." It was not until 1924 that a complete picture arose with Louis de Broglie's matter-wave speculation, and the genuine significance of quantum mechanics turned out to be clear. The absolute most unmistakable researchers consequently contribute during the 1920s to what is presently called the "new quantum mechanics" or "new physical science" Erwin Schrödinger, Wolfgang Pauli, Younger Heisenberg, Paul Dirac, and Max Born.

Afterward, the field was extended with work by Richard Feynman, Sin-Itiro Tomonaga, and Julian Schwinger for Quantum Electrodynamics improvement in 1947 and by Murray Gell-Mann for Quantum Chromodynamics

advancement. The obstruction that produces shaded groups on air pockets can't be clarified by a model that portrays light as a molecule. You may be clarified by a model that portrays it as a wave. These drawing demonstrations sine waves that look like waves outside of water being manifested from the two surfaces of the film of changing the width, yet that portrayal of light's wave idea is just an unrefined similarity. Early specialists contrasted in their clarifications of the principal idea of what You currently call electromagnetic radiation.

Some kept up that light and different frequencies of electromagnetic radiation are made from particles, while others affirmed that electromagnetic radiation is a wave marvel. In old-style physical science, these thoughts are commonly conflicting. Since the beginning of QM, researchers have recognized that neither thought without help from anyone can clarify electromagnetic radiation. Notwithstanding the achievement of quantum mechanics, it has some dubious components. For instance, the conduct of minuscule items depicted in quantum mechanics is altogether different from your ordinary experience, which may incite some level of skepticism.

A large portion of old-style material science is presently perceived to be made of uncommon instances of quantum physical science hypothesis as relativity hypothesis. Dirac presented the relativity hypothesis as a powerful influence for quantum material science to appropriately manage occasions that happen for a generous portion of the speed of light. Traditional material

science, nonetheless, additionally manages mass fascination (gravity), and nobody has yet had the option to carry gravity into a bound-together hypothesis with the relativized quantum hypothesis.

1.1 Explain the difference between quantum physics and classical physics?

1. Classical science is causal; absolute knowledge of the context helps the future to be computed. Similarly, absolute comprehension of the future makes

it possible to compute the past reliably. (To this statement, chaos theory is irrelevant; it speaks about how much you can do with insufficient knowledge.)

In quantum mechanics, not so. Objects are hardly particles or waves in quantum physics; they're a curious mix of both. You can only render probabilistic projections of the future, given full awareness of the past.

Two explosives with similar fuses will concurrently blast in classical physics. Two similar radioactive atoms will usually fuse at slightly different periods in quantum mechanics. On average, two equivalent uranium-238 atoms can experience nuclear decay isolated by billions of years because they are the same.

There is a rule sometimes used by scientists to distinguish classical physics from quantum physics. If in the calculations, Planck's constant occurs, it is quantum mechanics. It is classical physics if it doesn't.

Many scientists agree that quantum mechanics, while certain aspects have yet to be found, is the correct theory. It is necessary to derive classical physics from quantum physics to the degree that quantum properties are concealed. The reality is termed the "correspondence principle"

2. The movement that overthrew conventional mechanics was quantum physics. It is like establishing the distinction between the Bolsheviks and the Tsars to explain the disparity between them. Where should you even start?

They have a Newtonian picture of a clockwork world on one side. In this model, all observable life is a massive clock that ticks forward in time and, according to deterministic rules, shifts its structure predictably. According to a limited collection of basic mathematical rules, Newton saw his deity as a scientist who created the universe out of physical

components, setting them in motion. All the complexities and variety of natural events are essentially accountable for these rules. Likewise, all phenomena, no matter how complicated, can be interpreted in terms of these basic rules. "All discord, "harmony not heard", "harmony not understood."

On the other side, you have the quantum world, which appears to represent more like a slot machine than a clock from your point of view. You see the machinery of the quantum world as essentially probabilistic. It is unavailable to the experimenter whether there is unity behind quantum discord.

The quantum revolution goes far further than just incorporating chance as a basic attribute. It removes the Newtonian clock, exchanging it with a completely alien unit constructed of far more complex mathematics. The quantum revolution shows us that it is not just false; the classical viewpoint is essentially unsalvageable.

Let's begin to explore those Newtonian elements to be tossed into the garbage:

1. In all stages, particles & fields hold. You'll describe dynamic variables. The quantities used to characterize an object's motion, like position, momentum, velocity, and electricity, are dynamic variables. Classical physics claims that a system's dynamic parameters are clearly described and that they can be calculated to absolute precision. A traditional particle, for instance, occurs at a particular point in space at any specific moment in time and moves at a single velocity. And if the precise value of each variable is unknown, you presume that they occur and that only one value is taken.

2. Particles adopt predetermined trajectories as point-like artifacts. A particle is viewed in traditional mechanics as a dimensionless stage. By tracing a pathway through intermediate space, this point moves from A to B. As it moves around the surface, a billiard ball traces a straight line, a satellite in space traces an ellipse, & so on. The definition of a certain trajectory needs termed dynamic variables, & therefore the notion of a certain trajectory should be dismissed until the first argument above is abandoned.

3. As constant real numbers, complex variables. In classical mechanics, smoothly changing continuous parameters are dynamic variables. Quantum science derives its name from the finding that, in special conditions, certain amounts, most commonly energy & angular momentum, are constrained to such discrete 'quantized' values. The throughout values are Prohibited

4. Particles & waves are processes that are different. Classical mechanics has one particle structure and a related wave and field framework. This refers to the intuitive principle that, in opposite forms, the billiard ball as water wave passes from A to B. Nevertheless, these two processes are synthesized & treated in quantum mechanics within a single, marvelous structure. All physical objects are wave/ Particle hybrids.

5. Second Rule by Newton. Even without the four above-listed kinematic characteristics, $\sum F=ma \sum F=ma$ is more than false, and it isn't very meaningful. It is important to create a fundamentally new dynamic that is controlled by a somewhat different motion equation.

6. Predictability of the effects of measurements. The effects of measurements can be expected perfectly in classical physics, assuming a complete understanding of the method beforehand. In quantum mechanics, the effects of such experiments would be difficult to foresee, even though you have a complete understanding of a model.

It is no surprise because quantum mechanics took many decades to establish a multinational partnership with the above list in mind. Without these functions, how can you construct a coherent model of the universe? Luckily, it was not completely appropriate to scrap anything from classical physics. The rules of conservation are maintained. As fine as classical physics, Quantum Physics retains momentum, electric charge, and energy.

Although Newton's version of classical mechanics is discarded, conservation laws enable one to modify instruments from classical mechanics' more mathematically elegant Hamiltonian & Lagrangian formulations. The Hamiltonian formalism, which led to his eponymous equation, Erwin Schrodinger, decided to adopt. Lagrangian mechanics is adapted by Richard Feynman, which contributed to his direction of the integral formulation. Heisenberg created his original esoteric program known matrix mechanics.

These three methods to quantum mechanics (more of three, but there are standard formulations) are mathematically similar and effective. The formulation of quantum mechanics by Schrodinger is generally the one

that anyone encounters first, and it's the most applied formalism in the area. Now let's go back to the above list and swap Newton's parts with Schrodinger's parts:

1. At all stages, particles have wave function $\Psi(x,t)\Psi(x,t)$. For each moment in time, the wave function grants each point in space a complex number. This feature includes all usable particle knowledge inside it. Everything that can be learned regarding the particle's motion is derived from $\Psi(x,t)\Psi(x,t)$. You utilize Born's law to retrieve dynamic knowledge and measure $\Psi\Psi*\Psi\Psi*$ to get probability density of the direction of the particle, & you calculate $\phi\phi*\phi\phi*$ to get the average values of the momentum of the particle since $\phi(p,t)\phi(p,t)$ is the wave function's Fourier transform. This approach to kinematics is fundamentally different from traditional mechanics, which defines particles by listing dynamic variables' values

2. For wave function development, trajectories are substituted. If the wave function varies in time, the probability of the particle's detection of specific locations and moments changes. The equation of evolution is the Schrodinger equation that relies on time: $i\hbar\Psi'(x,t)=H\Psi(x,t) i\hbar\Psi'(x,t)=H\Psi(x,t)$. HH is a Hamiltonian machine operator, the self-adjoint operator equivalent to the cumulative system energy (explain below in 3rd point)

3. Hermitian matrices are complex variables. Schrodinger uses set Hermitian matrices or self-adjoint operators instead of real-valued, constantly changing dynamic variables to describe measurable quantities. Each measurable has a corresponding operator/matrix, such as momentum, position, energy, etc. The matrix/own operator's values decide the permitted values of corresponding observables. The atoms' energy ranges, for example, are the Hamiltonian operator's principles. This is another fundamental deviation from how classical mechanics deals with motion

4. Particle & wave unification. The Schrodinger equation's statistical study shows that it would have wavelike solutions, and therefore particles move like

waves. This ensures. You shouldn't imagine particles bouncing across their world as small spheres. Through visualizing the wave feature, the nearest you will come to visualizing the particle is the wave function, as previously mentioned in the first point above, assigns each point in space a complex number. In time, this world of complex numbers grows. What feels like this evolution? It looks like a region of quickly spinning phasors if you are acquainted with phasors (best visualization). The area of phasors for a single particle, more precisely, appears like a screw that bends in the direction of motion

5. The Schrodinger time-dependent equation eliminates the second rule of Newton

6. It's a spontaneous calculation. You would always not be able to determine the results of measurements in general, particularly though you have a full understanding of the quantum system before calculation (i.e., you identify $\Psi(x,t)\Psi(x,t)$). The calculation result is probabilistic. The potential consequences are determined by the operator's values that you observe (look at point 3). The wave

function projection decides the likelihood of each outcome on the operator's vectors.

So, here is a description of what Schrodinger feels like in quantum mechanics. There may be various specifics of alternative formulas, but the gist remains the same.

Hopefully, now it is evident that there are vast variations in classical and quantum mechanics. The quantum revolution

is one of the 20th century's most stunning theoretical advances, and the consequences of the revolution are yet to be truly felt in several respects. Quantum computation, for instance, is one subset that has not yet materialized. Of necessity, the philosophical and technical consequences will begin to change the 21st century in remarkable ways.

1. There is an 'of principle determinism in classical mechanics. In theory, in some gas canister, if you only have N atoms of gas neon, and you understood the location & momentum of each one of them, you might thoroughly explain the past of all time.

That does not suggest you cannot use mathematical techniques or interpret the movements as random (you will need to note 6 N numbers as a feature of time to treat these deterministically!). And in traditional physics, such approaches are incredibly helpful. It only implies that, regardless of the observation method, precise explainable properties like location & momentum are observable to some precision.

Things such as electrons & atoms You're meant to be viewed strictly as particles in classical mechanics, and items including light & other sources of electromagnetic radiation You're strictly treated as waves. (In classical mechanics, it

turned out there are several kinds of stuff that occur with light & electrons that just cannot be fully explained.)

2. Orthodox mechanics. There is a particular position & momentum of every particle. The pool tables have a friction coefficient that is almost fully uniform, & collisions are roughly elastic. True, a few of the backspin tricks appear a bit freaky.

There are features like position & momentum in quantum mechanics which are Not observable towards any precision, irrespective of the observation process. There is a limit on how precisely you can calculate all at once, specifically in the context of position & momentum.

You should think about a particle as a wave that encodes the likelihood of a given calculation being made. The odds decide the potential observations, and they are not known "trajectory".

3. On the atomic scale & below, the difference becomes important. Huge macroscopic structures that include saying, maybe 7,000,000,000,000,000,000,000,000,000 atoms in it, such as you and me, can have differences owing to quantum instability, which is such a small fraction of them, for nearly all purposes, they may be easily regarded as

classical objects. Indeed, the wave formula correlated with the table or a pool ball or human body gives a wavelength that is so extremely small that, for these broad artifacts, the quantum equations imitate the traditional ones to a considerable degree of precision.

An experiment with double slits. A wave that hits the surface with two tiny openings nearby can interact with itself, creating fringes of disturbance. In this film, one at a time, electrons are shot at pair of slits. Electrons are electrons for sure. Yet, it appears like the electrons do not adopt a particular course and appear spontaneously. They shape interruption fringes when a ton has been transmitted.

4. When Newton formulated his gravity principle and the mechanics referred to as calculus, classical physics took shape. Newtonian mechanics are three-dimensional: depth, height, width, and energy fall in little lumps, in packets, because a single packet is a quantum, and the theories of Planck You're soon referred to as "quantum theory." Quanta will act like particles, and waves will behave like quanta. It sounds counter-intuitive, but all particles and waves may be light, and the differentiation relies essentially on how it is studied.

5. The disparity between Classical & Quantum Theory is immense.

a)   A body still takes the least direction of action in classical philosophy, and only one path remains. A particle almost often selects the least action path in Quantum theory and simultaneously chooses other least action paths.

b)   There are 9 boxes & 10 pigeons, so two pigeons can end up with at least one package. In Classical Philosophy, this is. In Quantum Physics, no such thing exists. Only from two boxes, you may transfer endless electrons.

c) In Classical Mechanics, you can evaluate the direction & velocity of a particle concurrently with great precision. Quantum Mechanics follows the Heisenberg Instability Theory.

d) For macroscopic particles, Traditional Physics is applicable. Quantum Mechanics applies to microparticles.

6. Here is an easy analogy.

Suppose you use a sponge ball to play squash. The first thing you want is to model the sponge ball's dynamics mathematically, even though you might integrate it into the machine design, and you want to invent the machine that can play it with you. A traditional model would be enough for this.

Now, if you'd like to substitute the sponge ball via an electron, the vintage sponge ball model will break apart.

First, before it hits your bat, there is no deterministic way to know the ball's position. Then there is a likelihood that even if you have it right, it will tunnel thru the bat. So, you've just begun with a long list of unseen phenomena in traditional mechanics. These phenomena are modeled in QM mathematics, and it explains why things are happening beautifully for probabilistic theory.

However, the problem is that the world appears like a much more unusual place with this new model. In a wave equation, the ball is not a ball anymore but an eigenvalue. It is nothing like the world with which You are familiar. This publishes interesting puzzles about what mathematics implies. Visualizing is both mind-bending and confusing, yet so fascinating because it is very counter-intuitive.

Chapter 2: Father of the Quantum Entanglement

The world of physics undergone two significant revolutions around the turn of the 20th century, approximately at about the same time. The first one was the General Theory of Relativity by Einstein and was dealing with physics' fundamental field. The next was Quantum Theory, which indicated that energy existed as separate packages, each called a "quantum." This modern physics division enabled scientists across the subatomic domain to explain the relationship between energy & matter.

Einstein viewed Quantum Theory as a theoretical framework of explaining existence at an atomic stage, but he denied that it retained "a sufficient reference for all of physics." He felt it needed strong predictions backed by direct experiments to explain the fact. But it is not feasible to explicitly analyze specific quantum interactions, giving quantum physicists no alternative but to estimate the possibility that incidents will arise. Challenging Einstein, Quantum Theory was championed by Physicist Niels Bohr. The very act of partially studying the atomic domain alters the product of quantum interactions, he claimed. According to Bohr, probability-based quantum forecasts

correctly explain the truth.

For their work on quanta, Max Planck, and Niels Bohr, two of Quantum Theory forefathers, earned a Nobel Prize in Physics. In the theory of Photoelectric Effect, for which he received the 1921 Nobel Prize, Einstein is called the third father of the quantum theory since he specified light as quanta.

May 15, 1935: The (EPR) paper Einstein, Podolsk, & Rosen attempting to contradict Quantum Theory is published in the Physical Review.

Newspapers are quick to discuss with the public Einstein's distrust towards "new physics." "Einstein's essay "Could Physical Reality's Quantum-Mechanical Explanation Be Deemed Complete?" inspired Niels Bohr to publish a rebuttal. Modern studies, notwithstanding Einstein's protests, also upheld Quantum Theory. HOWEVER, the EPR presented topics that shape the basis for most of the physics science of today.

At the influential 1927 Solvay Meeting, which was participated by top scientists of the day, Einstein & Niels Bohr started to challenge Quantum Theory. Bohr was the victor from certain versions of this national discussion.

2.1 The Photon's Wave-Particle Duality

In 1807, Thomas Young, an English physicist in an experiment named Young's Interruption Experiment, asserted that light had a wave's characteristics. Young's interference study shows that lights (waves) going via two slits add or cancel one another together and then fringes of interference emerge. Since light is regarded as a wave, this effect cannot be clarified.

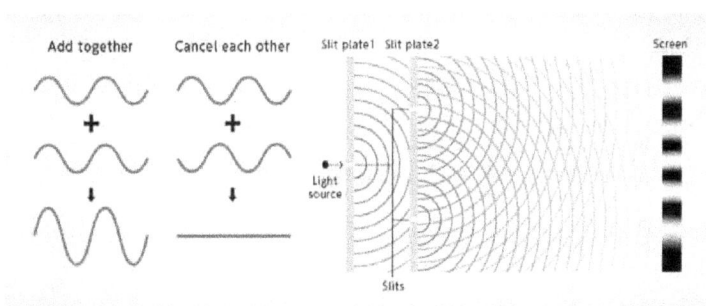

Young's interference fringes

2.2 The Light Is A Particle? Light Quantum Supposition Of Einstein -

Albert Einstein was indeed a scientist whose life encompassed Switzerland, Germany, and the United States. He succeeded in clarifying a photoelectric phenomenon in 1905, which was unexplainable if the light was called a wave. Einstein believed that a particle possessing energy proportional to its wavelength is electric.

The photoelectric impact is when the irradiation of blue light produces electrons on silicon. Red light, though, no matter how much time or even how strong the light is added, does not induce electron production from metal. You should conceive of photons as clusters of energy-containing particles to recognize this impact. Particles of high energy able with releasing electrons are blue light. Red light is a low-energy particle not able to produce electrons.

Thus, the light came to be named "photons or light quanta" because it has the characteristics of a wave and a molecule.

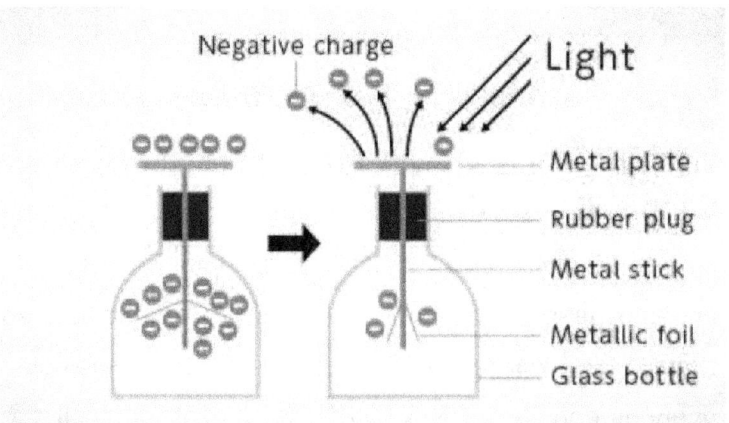

Photoelectric Effect Example: Leaf Electroscope

2.3 The Duality Of Photons

"Light not only is a wave, but it's a particle too." To improve your comprehension of this principle, an experiment was conducted.

This is Young's Double-slit Intrusion Trial or Interference Experiment. Using technologies to track single light particles, this analysis was performed to examine if interference fringes arise even though the light is significantly weakened to just one photon level. The experiment findings revealed that one Photon displayed an interference fringe.

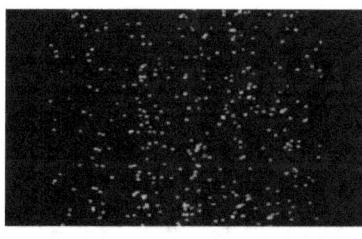 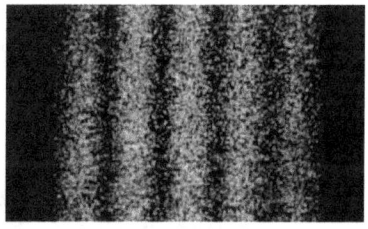

It functions as a particle when seen on the left as light is spread to an excessive brightness maximum and projected on a computer is observed. However, when the observed particle count rises, the interference fringe emerges, as shown on the right. From this, one can see the light still functions as a wave.

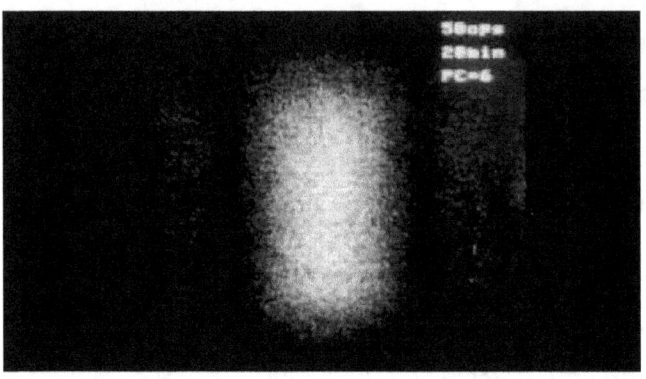

No interference fringe appears when one of the slits is closed.

Because one among two slits in this study is locked such that only a single-photon particle can travel into the other slit, then there is no fringe of disturbance. This revealed that this photon particle concurrently Count through two slits in the double slit interaction experiment and interacted independently.

> To see the video visit this link: www.bit.ly/videoqp

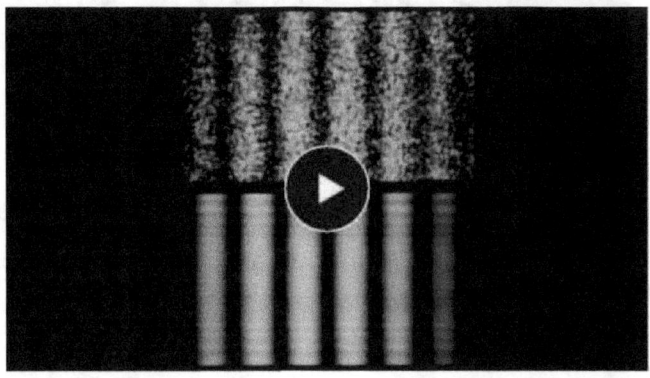

Interference Experiment of Young through single photon
(Hamamatsu Photonics, 1982)

These observations demonstrate that when a photon was identified as possessing the particle's properties while going through the double slit concurrently, interference looked like that of a stream, showing that perhaps the Photon does have dual attributes of a particle & a wave.

For about the first time in the world, this experiment caught the dual essence of Photon from a special camera.

2.4 The Nature Of The Photon

Based on Einstein's light quantum theory, quantum-mechanical studies and research verified the dichotomy of the Photon. In areas relating to the substance's contact and the light, which is received and released, the Photon now is known as a molecule and is thought to be a wave in regions related to the propagation of light.

You'll then, could it be that the Photon exhibits two opposite properties at the same time? Anyway, what is a photon?

It is understood that the Photon helps to transmit electromagnetic energy comprising the universe. The gravitational payout, heavy force, and force are the three other forces. In the physical universe in which it works, the Photon plays a major role and is profoundly involved with the origins of matter and life.

You will use light more reliably by understanding the essence of photon and build a modern revolutionary society that exceeds your intuition. That with the new sensitivity, creativity, and excitement of those who view this page, you might make this possible, then that will be fantastic.

2.5 "Photon on trial"

The "Koshi/Mitsuko no Saiban Photon on trial" is a tale set up about photons' dual existence, a mystical court drama. It was published in 1949 by Shinichiro Tomonaga, Nobel Prize laureate in physics. This work is a popular science essay that explains quantum mechanics understandably and has been read up to date by many people.

Recently, the thesis required estimating which of the two slits through which the photon passed, or in other terms, measurements and tests were done with the "You value," You're often performed. The analysis that searches for the essence of a photon is attracting further interest.

The 'Photon on trial' can be contained in a variety of quantum mechanics books. Those involved in more reading are urged to read the following.

- "Ryoshi rikigakuteki Sekaizou "The World is seen through Quantum Mechanics" Shinichiro Tomonaga, (Koubundou, 1965)

- "Kagami no Naka no Butsurigaku Physics in the Mirror" Shinichiro Tomonaga, (Kodansha Gakujutsu Bunko, 1976)

- "Ryoshi-rikigaku to Watashi (Quantum Mechanics & Me)" Shinichiro Tomonaga, Hiroshi Ezawa (editor) (Iwanami Bunko, 1997)

Nikkei Science's Photo courtesy
January 2014 problem of the Nikkei Science Again,
"Photon on trial" "does Mitsuko Namino naive?"

2.6 "Invitation to the photon."

"God says, 'Let the light be here,'..."

People have assumed that light was with Heaven or that light was there with God since ancient times. Human beings avoid their heads in the Sun, the creator of all sorts of lives, and see him as God.

Hamamatsu Photonics created an instructional video about light in 1984.

Chapter 3: Young's Double Slit Experiment

Isaac Newton didn't, whereas Christiaan Huygens assumed that light was a wave. Newton thought that there are many other reasons for color and the results of interference & diffraction that You appear at the time. His opinion usually prevailed owing to Newton's enormous stature. The fact that the theory of Huygens worked was not deemed proof that was conclusive enough to show that a wave is light. Many years after, when the English physician & scientist Thomas Young (1773-1829) performed his now-traditional double-slit experiment in 1801, the confirmation of the wave nature of light came several years later (see Figure).

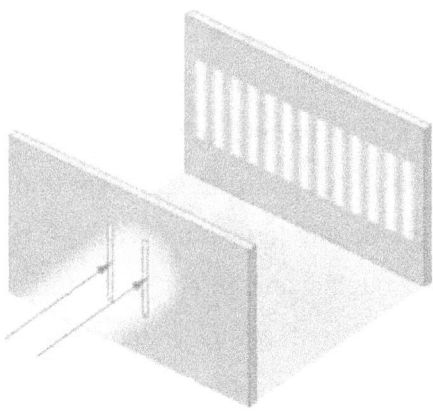

Double slit experiment by Young. Here, pure wavelength light transmitted via a pair of vertical slits diffracts various vertical lines stretched horizontally into pattern upon this screen. The light will create two lines on the panel without diffraction interruption.

Why don't You normally encounter wave behavior light, as found in Young's double-slit experiment? 1st, to demonstrate pronounced wave impact, light must communicate with anything short, such as the tightly spaced slits utilized by Young. Furthermore, Young initially transmitted light from a source, the Sun, though some single slit allowed the light to be rather coherent. By consistent, it means that waves are also in phase or to have some clear association with phase. Incoherent implies spontaneous phase relationships between the waves. Why would Young then move a double slit across the light? This query would respond that two-slits give two coherent light sources that then interact destructively or constructively. Young, utilized sunlight, in which each wavelength shapes its template, making it harder to see the effect. To explain the effect, it is demonstrated the double-slit experiment along with monochromatic light (single λ). The pure positive and disruptive interference of these two waves possessing a similar wavelength & amplitude, as seen in figure.

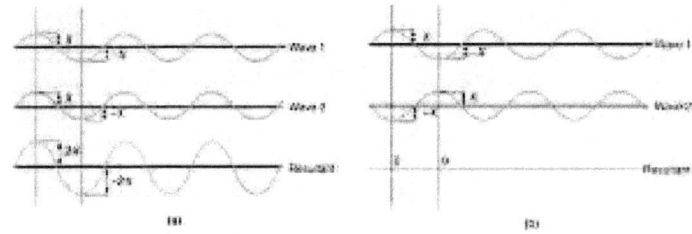

The wave amplitudes sum up. (a) When similar waves are in motion, pure positive interference is produced. (b) Pure disruptive interruption happens where precisely the same waves are out of alignment, or half a wavelength is moved.

It diffracts via semicircular waves as light flows via narrow slits, as seen in Figure a. If waves are peak to trough-to-trough or crest, pure positive interference happens. There is pure disruptive intrusion when they are crest-to-trough. For you to see a pattern, the light should fall on a projector and be scattered through your eyes. Figure b shows an analogous sequence with water waves. Notice that regions of positive and disruptive interference travel out from slits at demarcated angles to the initial beam. As you can see below, such angles rely on wavelength & the space between the slits.

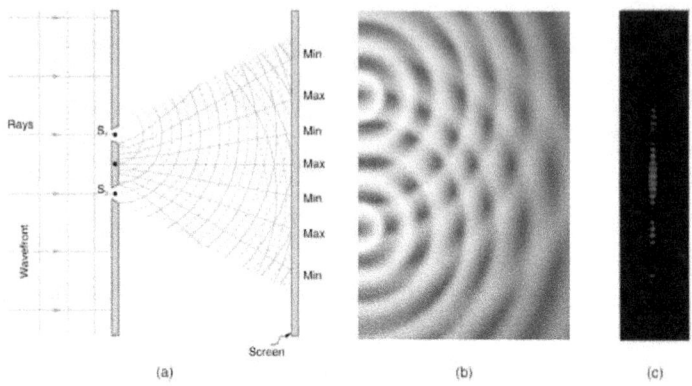

Double slits create two coherent influences of waves that interact. (a) From each break, the light expands (diffracts), therefore slits are narrow. Such waves converge and intervene constructively and destructively. And if the light comes onto a projector and is dispersed through your eyes, you will perceive anything. (b) For water waves, the double-slit interference pattern is almost like that of light. Wave action is highest in positive interference regions and least in disruptive interference regions. (c) You also see a pattern like this when light penetrates through the double-slit drops on a screen. (with credit: PASCO)

As shown in Figure, it will be considered how two waves pass from slits to screen for explaining the double-slit interference trend. On the screen, each slit is indeed a different length from a particular moment. Therefore, numerous wavelength numbers fit in each direction. Waves originate from slits the phase (crest to crest) but can wind up out of the phase (crest-to-trough) on the screen if paths vary by half wavelength in duration, intervening destructively, as seen in Figure a. If the paths vary greatly by a complete wavelength, the waves turn up at the screen in phase (crest-to-crest), tampering constructively, as shown in Figure b. More normally, if any half-integral quantity of wavelengths [(3/2) λ, (1/2) λ, (5/2) λ,] differ in paths taken by two waves, then devastating interference happens. Similarly, where some integral quantity of wavelengths, i.e., λ, 2λ, 3λ, vary in the directions followed by two waves, positive intervention exists.

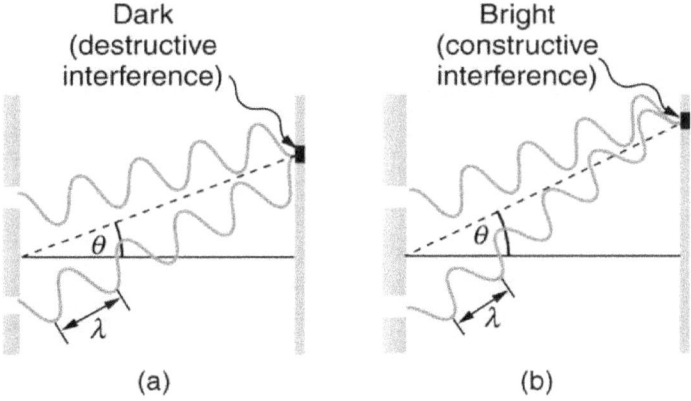

(a) (b)

From slits to a focal point on a computer, waves take various routes. (a) Since one direction is half wavelength longer than the other, disruptive interaction exists here. In

the process, the waves begin but appear out of phase. (b) the Constructive interaction happens here when one direction is longer than the other by a whole wavelength. The waves begin & appear in phase.

3.1 Take-Home Experiment: Using Fingers As Slits

Look at a lamp, like an incandescent bulb or streetlamp, through a restrictive gap between the two fingers that are kept close to each other. What sort of pattern are you seeing? When you permit the fingers to transition a little further apart, how would it change? Like yellow light from either sodium vapor lamp, it is more distinct than just an incandescent bulb for monochromatic origin.

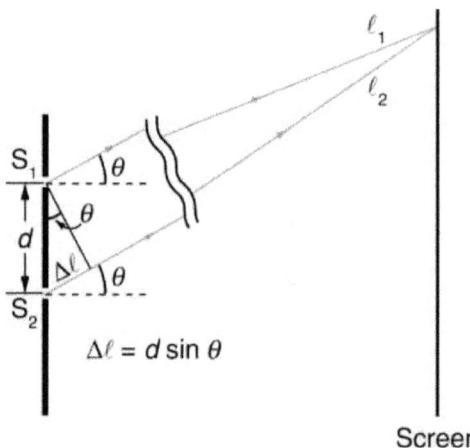

The paths on the screen from each slit to the common point vary by a quantity of dsinθ, considering that the screen distance is far longer than the length among slits.

For waves flowing from two slits towards a popular point on a screen, Figure 5 illustrates how to calculate the path length discrepancy. Compared to the gap between the slits, if the screen is a great distance apart, then the angle between the route and a line from slits to screen is about the same with each path (see the figure). The differentiation between paths is seen in the figure; basic trigonometry is seen as d sin θ, where d is the difference between slits. The direction length gap would have to be an integral multiple of the wavelength to achieve constructive interaction for a double slit, or d sin θ = mλ, for m = 0, 1, −1, 2, −2, . (i.e., constructive).

Similarly, for a double slit to achieve disruptive interference, the path length gap must be a half-integral multiple wavelength.

dsinθ=(m+12) λ,

for m=0,1,−1,2,−2,... (destructive)dsin☐θ=(m+12)λ, for m=0,1,−1,2,−2,... (destructive),

Where λ is light's wavelength, d is the difference between slits, & θ, as discussed above, is the angle from the original orientation of the laser. The order of the interference is

called m. For instance, fourth-order interference is m = 4.

The double-slit interaction equations suggest the creation of a sequence of bright & dark lines. The light stretches out horizontally to each side of the incident beam for vertical slits through a pattern known as interference fringes, seen in Figure 6. On either side, the color of the bright fringes drops off, becoming darkest at the middle. The nearer slits are, the farther the bright fringes reach away. You can see it by analyzing the d sin = mλ equation, for m = 0, 1, −1, 2, −2,

The layout d is for set λ & m, the larger must be θ, as sin=mλ dsin☐θ=mλd. This is correlated with the argument that when the target the wave meets is thin, wave impacts are more visible. Small d gives huge θ, so a major impact.

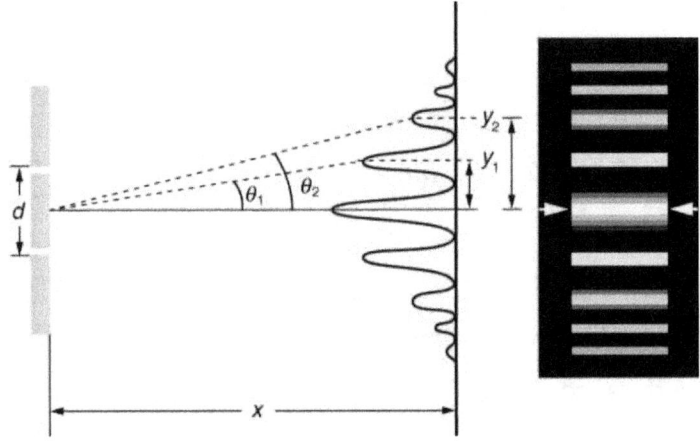

An amplitude that drops off with angle has the interaction pattern for the double slit. The picture displays many dark & bright lines, or fringes, produced by light flowing through the double slit.

3.2 Example 1. Finding A Wavelength From An Interference Pattern

Suppose you transfer light via two slits apart by 0.0100 mm from the He-Ne laser & notice that the third bright line is shaped at the angle of 10.95o compared to the incident beam on a panel. What's the wavelength of light?

Strategy

The 3rd bright line is attributed to positive intervention in the third magnitude, indicating that m = 3. d = 0.0100 mm & θ = 10.95° are given to you. The wavelength can therefore be found for positive interference utilizing equation d sin θ = mλ.

Explanation

This equation is DsinΘ = M Λ. Resolving for wavelength Λ provides λ= dsinθm λ = dsinθ m.

Substituting known values yields

λ=(0.0100 nm)(sin10.95◦)3 =6.33×10−4 nm=633 nmλ=(0.0100 nm)(sin☐10.95◦)3 =6.33×10−4 nm=633 nm

Discussion

This is the wavelength of light produced by the popular He-Ne laser with three digits. The red color is, not coincidentally, identical to that produced from neon lights. More significant, though, is the reality that wavelength detection disturbance patterns may be used. Young performed this with visible wavelengths. This empirical approach is also commonly used for electromagnetic spectrum analysis. Angle for constructive interaction increases with λ for a given order, such that spectra (intensity versus wavelength measurements) can be produced.

3.3 Example 2. Evaluating High Ranking Order Possible

Patterns of Interference, as there's a limit on how large m can be, don't have an endless no. of lines. What's the maximum potential constructive interference with the method mentioned in the previous example?

Strategy and Concept

Constructive intervention is defined by the equation d sin θ = mλ (for m = 0, 1, −1, 2, −2, . . .). The greater m is, the larger sin θ is, for set values of d & λ. The highest worth sin θ could have, therefore, is 1, with an angle of 90°. (Larger angles mean the light goes backward & does not at all hit the screen.) Let us discover whether m correlates to this maximal angle of diffraction.

Solution

Solving equation $D \sin \Theta = M \Lambda$ for M gives $\lambda = d \sin \theta m$ $\lambda = d \sin \theta m$.

Taking $\sin \Theta = 1$ and substituting values of D & Λ from the previous example gives

m= (0.0100 mm) (1)633 nm≈15.8m= (0.0100 mm) (1)633 nm≈15.8

Consequently, major integer M can be is 15, or M = 15.

Discussion

The number of fringes is based on the distinction of wavelengths and slits. For large separations of slits, the number of fringes would be quite large. The size of the interference pattern varies, though, if slit spacing becomes much wider than a wavelength, such that the screen has 2 bright lines formed by slits, as predicted as light acts as a beam. You also remember that farther from the middle, fringes get fainter. Therefore, not even 15 fringes

can be detected.

3.4 Section Summary

- The double-slit experiment by Young provided conclusive proof of light's wave character.

- The superposition of the light by two slits obtains an interference pattern.

- While $d \sin \theta = m\lambda$ (for m= 0, 1, −1, 2, −2,.), there's a constructive interference where d is the gap between the slits, θ is angle proportional to the direction of the incident, and m is an order of interfering.

- While $d \sin\theta = m\lambda$ (for m = 0, 1, −1, 2, −2, . . .) there is disruptive intervention.

3.5 Conceptual Questions

1. A solo light beam is split into two sources by Young's double-slit experiment. Will the same pattern be acquired for two separate light sources, like the headlights of a distant vehicle? Just explain.

2. To conduct Young's double-slit experiment in air, assume you utilized the similar double-slit and then replicated the water experiment. Do the angles get greater or smaller with the same portions of the interruption pattern? Will the light's color change? Just explain.

3. Can a condition be generated in which only disruptive interference exists? Just explain.

4. For a red light's clear wavelength projected onto a double slit, Figure 7 indicates the central part of the interference sequence. Currently, the pattern is a mixture of single-slit & double-slit interference. Notice that light spots are distributed uniformly. Is this a function of double-slits or single-slits? Notice that on either side of the middle, some of the bright points are dim. Is this a single-slit or double-slit feature? What is smaller, the width of the slit or the division between the slits?

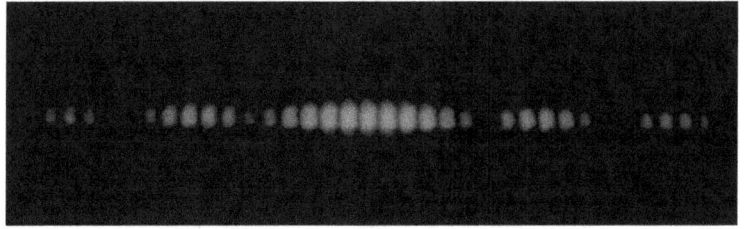

This trend of double-slit interference reveals symptoms of single-slit interference

Chapter 4: What is Heisenberg's Uncertainty Principle?

The Heisenberg principle of uncertainty notes that both the position & momentum of some objects cannot be determined or estimated correctly. This argument is grounded on the duality of matter in the wave-particle. While Heisenberg's principle of uncertainty in the macroscopic universe may be overlooked (uncertainties in the velocity & position of objects with comparatively large masses are insignificant), in the quantum world, it has considerable importance. As atoms & subatomic particles possess very low masses, any improvement in their location precision would be followed by an increase in their velocity-related uncertainty.

In quantum mechanics, the uncertainty principle of Heisenberg is a basic theory that describes why it is difficult to concurrently calculate more than one quantum component. Another consequence of the theory of complexity is that it is difficult to reliably estimate a system's energy in a short time.

4.1 Why is calculating both momentum & position concurrently impossible?

Consider an illustration where the position of an electron is determined to demonstrate Heisenberg's uncertainty theory. A photon must interfere with it & return to the measurement system to determine an entity's location. Because photons carry some limited momentum, as the photon interacts with an electron, a transition of moments may occur. This transition of moments would allow the electron's momentum to increase. Any effort to quantify a particle's location would raise confusion regarding the magnitude of its momentum.

Trying to apply a similar illustration to a macroscopic entity, it can be observed that the theory of uncertainty of Heisenberg has a marginal effect on macroscopic world measurements. There would also be a transference of momentum from photons to balls when measuring a basketball's location—the mass of a photon is somewhat less than the mass of the ball, though. Any momentum regarded as a separate ball by photon may also be ignored.

The theory of Heisenberg uncertainty puts a constraint on the precision of simultaneous momentum and position measurements. The more accurate position measurements are, the less precise momentum measurements would be, & vice versa. The physical basis of the theory of Heisenberg uncertainty resides in the quantum method. By conducting a calculation on the device, evaluating the location disturbs it enough to render the assessment of q imprecise & vice versa. Below, you will read in-depth regarding the theory.

The Heisenberg Uncertainty Theory states that this will not precisely determine both velocity & position at the same time for particles that exhibit both natures of particle & wave. The method is called after Younger Heisenberg, the German physicist who formulated uncertainty in 1927. As Heisenberg was attempting to construct a model of intuitive quantum physics, this theory was developed. He found that in interpreting such numbers, there You're certain underlying forces that constrained the actions.

This theory fundamentally stresses that there would be an error in the simultaneous calculation of position and momentum or the velocity of microscopic matter waves in such a way that the result of an error in momentum and position measurement is equal to or greater than that of an integral multiple of constant.

4.2 Heisenberg Uncertainty Principle Formula and Application

If Δx is a measurement of position error & Δp is momentum measurement error, then

$$\Delta X \times \Delta p \geq \frac{h}{4\pi}$$

Meanwhile, momentum, p = mv, uncertainty principle Heisenberg formula can be instead written as-

$$\Delta X \times \Delta mv \geq \frac{h}{4\pi} \text{ or } \Delta X \times \Delta m \times \Delta v \geq \frac{h}{4\pi}$$

Where ΔV is an error in velocity measurement & presuming persistent mass constant during the experiment,

$$\Delta X \times \Delta V \geq \frac{h}{4\pi m}.$$

Precise position or momentum measurement inevitably suggests larger uncertainty/error in the other quantity measurement.

Directing the Heisenberg theory to the electron in the orbit of the atom, along h=6.626 ×10^{-34} Js & m=9.11×10^{-31}Kg,

$$\Delta X \times \Delta V \geq \frac{6.626 \times 10^{-34}}{4 \times 3.14 \times 9.11 \times 10^{-31}} = 10^{-4} \text{ m}^2 \text{ s}^{-1}$$

If the electron's location is determined accurately to its size

(10-10 m), the velocity calculation error would be equivalent to or greater than 106 m or 1000 km.

The Heisenberg theorem refers only to microscopic particles of dual type & not to a macroscopic particle of a very small wave nature.

4.3 Explaining Heisenberg Uncertainty Principle with An Example

Electromagnetic radiation & waves in microscopic matter show a dual character of mass dynamism & wave existence. The direction and velocity/momentum of waves of macroscopic matter can be precisely measured simultaneously. For starters, it is possible to calculate the direction and speed of a moving vehicle at the same moment, with minimal error. In microscopic particles, though, it would not be practical to set the location and simultaneously measure the particle's velocity/momentum.

An electron has a density of 9.91×10^{-31} Kg in an atom. Naked eyes won't see tiny particles like that. The electron can collide with a powerful light & illuminate it. Illumination allows to define the electron's location and to Young it. Although aiding in detection, the strong light source's collision increases the electron's velocity and lets it travel away from the original location. Thus, the velocity/momentum of the object will have varied from the original value before establishing the location.

Therefore, deviations arise in the calculation of velocity or momentum while the location is accurate. In the same way, the momentum calculation would correctly modify the location.

Hence, any momentum or position can only be calculated correctly at some point in time.

In both position and velocity, simultaneous calculation of both would have a mistake. The flaw in calculating both location and momentum was quantified at the same time by Heisenberg.

4.4 Heisenberg's γ-ray Microscope

Gamma-ray microscope of Bohr/Heisenberg is a striking thought experiment demonstrating the confusion principle. You shine this along with the light wave of the wavelength λ to observe an atom, claim an electron, and gather Compton dispersed light in the microscope purpose whose circumference subtends the angle with an electron, as seen in below figure

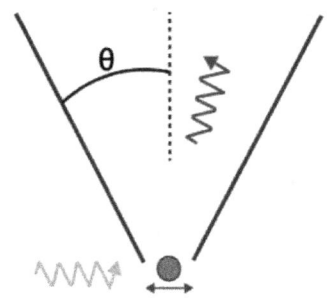

Delta x, the accuracy with which electron may be found, is determined by the microscope's resolving payout,

$\sin \theta = \frac{\lambda}{\Delta x} \Rightarrow \Delta x = \frac{\lambda}{\sin \theta}$ $sin=\Delta x \lambda \Rightarrow \Delta x = sin\theta\lambda$

It seems that Delta x can be rendered as small as needed by having λ small, which is why You chose γ-ray & by keeping sin θ high. But, due to the theory of confusion, you may do so only at the cost of your understanding of the x-component of the momentum of an electron.

The photon must remain in the angled cone & so its x component of momentum will differ within ±(h/λ) sin θ to detect the Compton dispersed photon via the microscope. This suggests that the extent of the electron's rebound momentum is unknown.

$$\Delta p_{x}=\frac{2h}{\lambda}\sin\theta \Delta px = \lambda 2h\sin$$

The result of uncertainty produces,

$$\Delta x \Delta p_{x} = \frac{\lambda}{\sin\theta}\frac{2h}{\lambda}\sin\theta = 4\pi h \Delta x \Delta px = \sin\theta\lambda\lambda 2h\sin\theta = 4\pi h$$

4.5 Does Uncertainty Principle of Heisenberg Noticeable in All the Matter Waves?

Heisenberg's theory extends to all the matter waves. The calculation uncertainty of just about any of the 2 conjugate factors, whose measurements happen for being joule sec, would be driven by the Heisenberg's value as position momentum, time energy.

However, it can only be visible and important for tiny particles, such as an electron with a very low density. The error is very tiny and marginal with a larger particle with a heavy mass.

4.6 Heisenberg Uncertainty Principle Equations

The theory of uncertainty of Heisenberg may be viewed as a rather accurate mathematical concept that explains quantum systems' existence. So, you will also consider 2 common equations connected with the theory of uncertainty. They remain.

Equation 1: $\Delta X \cdot \Delta p \sim \hbar$

Equation 2: $\Delta E \cdot \Delta t \sim \hbar$

Where,

$\hbar$ = value of Planck's constant divided by 2*pi
ΔX = uncertainty in position
Δp = uncertainty in the momentum
ΔE = uncertainty in energy
Δt = uncertainty in the measurement of time

4.7 Explained Numerical Difficulties on Uncertainty Principle of Heisenberg

1. Calculate the variance of momentum of electron if position of electron is determined at a precision of + 0.002 nm. If the electron's momentum is h / 4pm, 0.05 nm, the concept of this magnitude is troublesome.

a) $\Delta x = 2\times 10^{-12}$ m; $\Delta X \times \Delta mV \geq \frac{h}{4\pi} $ }$4\pi h$
= \frac{6.626\times   {{10}^{-34}}}{4\times 3.14}4×3.146.626×10−34

∴ $\Delta mV \geq \frac{h}{4\pi}$ \Delta   x}$4\pi\Delta xh$
≥ \frac{6.626\times   {{10}^{-34}}}{4\times   3.14\times 2\times   {{10}^{-12}}}\,4×3.14×2×10−126.626×10−34 = 2.64 ×

10^{-23} Kg m s^{-1}

b) Momentum $mv = \frac{h \times 5 \times 10^{-11}}{4 \times 10^{-12}} = \frac{6.626 \times 10^{-34} \times 5 \times 10^{-11}}{4 \times 10^{-12}}$, $4 \times 10^{-12} h \times 5 \times 10^{-11}$

$= 4 \times 10^{-12} 6.626 \times 10^{-34} \times 5 \times 10^{-11} = 28 \times 10^{-33}$

Error in the measurement of momentum is 10^{10} times greater than actual momentum. The specified momentum would not be satisfactory.

2. A maximal error of 1μm may be measured by the location of the chloride ion on the substance. If chloride ion's mass is 5.86x10-26Kg, what is the mistake in its calculation of velocity??

$\Delta x = 10^{-6}$ m; $\Delta X \times \Delta mV \geq \frac{h}{4\pi} 4\pi h$

$= \frac{6.626 \times 10^{-34}}{4 \times 3.14} 4 \times 3.14 \, 6.626 \times 10^{-34} = 5.28 \times 10^{-35}$ Js

∴ $\Delta V \geq \frac{h}{4\pi m \Delta x} \ge \frac{6.626 \times 10^{-34}}{4 \times 3.14 \times 5.86 \times 10^{-26} \times 10^{-6}}$, $4\pi m \Delta x h$

$\geq 4 \times 3.14 \times 5.86 \times 10^{-26} \times 10^{-6} 6.626 \times 10^{-34} = 9 \times 10^{-4}$ m s^{-1}

3. The lifetime of the atom's excited state is 3 × 10-3s. What's minimum power uncertainty in eV?

Energy and time are conjugate pairs along Js unit. The result of measurement error is provided by the principle of Heisenberg.

$$\Delta t \times \Delta E \geq \frac{h}{4\pi} = \frac{6.626 \times 10^{-34}}{4 \times 3.14} = 5.28 \times 10^{-35} Js$$

Presuming the maximum error in measurement of lifetime equal to the lifetime = 3×10^{-3} s

$$\Delta E \geq \frac{h}{4\pi m \Delta x} = \frac{1}{3 \times 10^{-3}} \times 5.28 \times 10^{-35} J$$

∵ 1 Joule = 6.242×10^{18} ev,

Uncertainty in determination of the energy of atom = ΔE
= $6.22 \times 10^{18} \times \frac{1}{3 \times 10^{-3}} \times 5.28 \times 10^{-35}$

= 1.1×10^{-13}

4. A 10.1gm Yout Ball Youighing has 0.1g water on it. For a fixed velocity with an impulse uncertainty of 10-6 kg m/s, the ball travels. What is the confusion regarding the calculation of ball, water & electron location in water molecules?

$$\Delta X \times \Delta p \geq \frac{h}{4\pi}$$

Velocity being continuous, uncertainty in a measurement of momentum is associated with a mass of matter.

Uncertainty in momentum of dry ball = mass $\times 10^{-6}$ = $10 \times 10^{-3} \times 10^{-6}$ Kg m s^{-1}.

Uncertainty in the momentum of water = mass $\times 10^{-6}$ = $0.1 \times 10^{-3} \times 10^{-6}$ Kg m s^{-1}.

Uncertainty in the momentum of electron = mass $\times 10^{-6}$ = $9 \times 10^{-31} \times 10^{-6}$ Kg m s^{-1}.

Uncertainty in the position measurement is also inversely proportional to uncertainty in the momentum

$\Delta X \geq \frac{h}{4\pi m \Delta p} \alpha \frac{1}{\Delta p}$

$\Delta X_b : \Delta X_w : \Delta X_e = \frac{1}{10^{-8}} : \frac{1}{10^{-10}} : \frac{1}{9 \times 10^{-37}}$, $10-81:10-101:9 \times 10-371$

= $10^8 : 10^{10} : 1.1 \times 10^{36}$ or

= 1 : 10^2 : 10^{28}

5. Certain minimal uncertainty in electron locations if their velocities are established with an accuracy of 3.0-10-3m/s?

Solution:

$\Delta u = 3.0 \times 10^{-3}$ m/s

Uncertain momentum $\Delta p = m \Delta u$

Uncertainty in position $\Delta x = \hbar/(2\Delta p)$

For electron

$\Delta p = m \Delta u$

$= (9.1 \times 10^{-31}$ kg $\times 3.0 \times 10^{-3})$

$\Delta p = 2.73 \times 10^{-33}$ kg.m/s

\Delta x = \frac{h}{\Delta p }$\Delta x = \Delta ph$

$\Delta x = 0.12$ m

4.8 Schrodinger's cat

A popular hypothetical experiment intended to find out a mistake in understanding superposition in Copenhagen as it relates to quantum theory is Schrödinger's cat.

This iteration of the simulated experiment is somewhat simplified:

A live cat, along with an ax, vial of hydrocyanic acid & very limited volume of radioactive material, is put into a steel chamber. Suppose even a single atom of a toxic material decays during the evaluation process. In that case, the hammer is caused by a relay device, which destroys the hazardous gas vial, which allows the cat to die.

This mental experiment was developed in 1935 by Nobel Prize-winning Erwin Schrödinger, an Austrian physicist, to

find a paradox between whether quantum theorists hold to

be valid regarding the origin and actions of matter at the microscopic level & what the average individual experiences to be true at macroscopic level with the unassisted human eye.

The character of the spectator in the quantum mechanics

The Copenhagen explanation of quantum mechanics, which was the dominant hypothesis, suggested that atoms or photons occur in several states that correlate with various potential outcomes & possibilities, called superpositions, do not agree to a definitive condition till they are detected.

Schrödinger's thinking experiment was intended to demonstrate what Copenhagen's explanation would look like if the scientific language used to describe superposition in the microscopic universe were substituted with macroscopic concepts the ordinary individual might imagine and comprehend. In the experiment, the observer cannot tell whether or not an atom of the material has decayed, and therefore, does not know whether the vial has burst, and the cat has been destroyed.

According to the quantum law, the cat would be both

dead & alive before anyone looks in the box under Copenhagen's view. In the jargon of quantum mechanics, the cat's tendency to be simultaneously dead & alive when examined is referred to as quantum indeterminacy or paradox of the observer. The rationale behind the paradox of the observer is the demonstrated potential of observation to affect results.

Schrödinger recognized that superposition exists; physicists can show its presence during his lifespan by researching disturbance in light waves. However, Schrödinger thought just whether the resolution of possibilities truly happens. The thought experiment was meant to question whether it was reasonable for observation to be a cause. Would a cat not be either alive or dead, even if it's not observed?

The cat example of Schrödinger has been used over the years to explain new hypotheses of how quantum mechanics functions. And In the Several Worlds view of quantum law, e.g., the cat is both dead and alive. The observer & the cat reside in two worlds in this view — one in which the cat is alive & the one in which the cat is dead.

What physicists have discovered about the existence of matter at the microscopic level and its connection to humans' experience at the macroscopic level has not yet been thoroughly investigated. The observer's position remains a significant topic in the analysis of quantum entanglement and is an endless subject of debate and hypothesis in quantum computing and pop culture. It is rumored that Schrödinger himself claimed, later stages of life, that he hoped he had never seen that cat.

4.9 This Twist Cat Paradox of the Schrödinger Has Main Insinuations for the Quantum Theory

A lab presentation of classic "Wigner's friend" thought experimentation could reverse cherished theories about truth

Credit: Getty Images

What would it sound like to be simultaneously living & dead?

This issue irked & influenced Eugene Wigner, a Hungarian-American physicist in the 1960s. He was annoyed by paradoxes resulting from quantum mechanic's vagaries the principle regulating microscopic world that implies, among several other counter-intuitive stuff, that unless the quantum system is examined, it does not contain definite properties. Consider his fellow scientist Erwin Schrödinger's famed thought experiment during which a cat is stuck in a box of poison that would be released when an atom (radioactive) decay.

Radioactivity is a quantum process, but when the package is opened, the tale goes, an atom has just decayed & not decayed, holding the unlucky cat in limbo—a so-called superposition among life & death. But does the cat feel like being in the superposition?

Wigner sharpened contradiction by considering a (human) associate of his shut in the lab, measuring the quantum device. He claimed it was ridiculous to suggest his friend lives in the superposition of getting seen & not seen a deterioration until & before Wigner opens the lab door. "'Wigner's friend' thought study demonstrates that items can get very odd if an observer is also being observed," says Nora Tischler at Griffith University, a quantum physicist in Brisbane, Australia.

Now Tischler & her collaborators also carried out a variant of Wigner's buddy examination. By integrating the traditional thought experiment with some other quantum head-scratcher named entanglement-a-mechanism that connects particles through large distances, they have also derived a new theorem, which they say places the best limits on the basic nature of existence. Their research, which appeared on Nature Physics on August 17, has consequences for the position that consciousness could play in quantum mechanics & also if the quantum theory could be substituted.

The latest thesis is an "essential step forward for an area of analytical metaphysics," claims Steinberg, a quantum physicist of the University of Toronto, who's not interested in research.

4.10 The Matter of The Taste

Before quantum mechanics comes along in the 1920s, scientists expected their hypotheses to be deterministic, producing forecasts for experiments with confidence. Yet quantum theory tends to be fundamentally probabilistic. The textbook version, sometimes known as "Copenhagen interpretation," says that once the system's properties are calculated, they will encompass countless values. This superposition often falls into a single-phase when the system is detected, & physicists will never accurately determine the phase. Wigner retained the prevalent opinion that awareness somehow causes a superposition to fall. Thus, his imaginary friend could discern some definitive result as she/he rendered a measuring system Wigner will never see him/her in superposition.

This opinion has now gone out of fashion. "People information of the quantum mechanics quickly reject

Wigner's perception as freaky or inadequately defined because it causes spectators unique," said Mr. Chalmers, a cognitive and philosopher at the " University of New York." Many physicists fully agree that inanimate artifacts will kick out quantum systems of superposition by a mechanism recognized as decoherence. Scientists attempting to modify complicated quantum superpositions in the lab will see their diligent work ruined by swift air particles interacting with their structures. All of them carried out their experiments at supercooled temperatures & aim to protect their frameworks from

vibrations.

Several conflicting quantum interpretations sprang up over decades that use less mysterious processes, such as decoherence, to clarify why superpositions disintegrate without invoking awareness. Most views take the far more extreme view that there's no failure at all. Everyone has their own strange & wonderful perspective on Wigner's examination. The most extreme is the "many worlds" concept, which states that if you create a quantum measurement, truth fractures, generating alternate dimensions to satisfy any conceivable result. Thus, Wigner's friend will break into two copies &, "with strong enough super-technology," he might still calculate the human to have been in superposition outside the lab, claims many-worlds fan and quantum physicist Lev Vaidman from Tel Aviv University.

The alternate "Bohmian" theory (called after physicist David Bohm) suggests that quantum structures do have specific features; You don't even know sufficient about such systems to specifically determine their behavior. In either scenario, the friend has a single encounter, but Wigner may always calculate the person to be in superposition due to his incompetence. In comparison, a new addition on the block named the QBism explanation accepts the probabilistic aspect of the quantum theory heartily (QBism, spelled "cubism," is simply short of quantum Bayesianism, the reference to the 18th-century mathematics expert Thomas Bayes's research on probability.) So, QBists claim that a person may only utilize quantum mechanics for calculating how to calibrate her/his assumptions about what she/he can test in an experiment. "Measurement results should be considered unique to an agent who produces these measurements," claims Ruediger Schack, one of QBism's members from Royal Holloway, of University of London. According to QBism's beliefs, quantum principles can't teach you anything regarding the fundamental condition of truth, nor will Wigner utilize it to comment about his friend's experiences.

Another interesting interpretation, named retro causality, causes incidents in the future to affect the past. "In the retro, causal account, a friend of Wigner did encounter something," states a physicist Ken Wharton from San-Jose-State University; he is also an enthusiast for such a time-twisting interpretation. But the "something" friend feels at the point of measurement will rely upon Wigner's preference of how one can observe the individual later.

The problem is that every understanding is similarly good—or bad—at predicting the results of quantum experiments, so deciding amongst them falls to taste. "No one recognizes what this solution seems to be," Steinberg notes. "You even don't know all the potential solutions currently present is exhaustive."

Some models, termed failure theories, can enable testable forecasts. These models add on a function that causes a quantum device to break as it becomes too big, explaining why only cats, humans, and other macroscopic artifacts can't be in the superposition. Experiments are ongoing to look for indications of these collapses, but then they have not discovered anything. Quantum physicists are now putting ever greater structures into superposition: A team, last year in Vienna, recorded doing so for 2,000 atom molecules. Many quantum interpretations suggest there is no excuse why these attempts to unsudden superpositions could not proceed upward indefinitely, presuming researchers will formulate the correct tests in sterile lab conditions such that decoherence can be prevented. However, collapse hypotheses posit that a cap would one day be hit, irrespective of how cautiously tests are planned. "If you manipulate and try a classical investigator human, treat it like a quantum system, this will collapse completely," says Angelo Bassi, promoter of collapse theories & quantum physicist at the University of Trieste present in Italy.

4.11 A Method to Observe Wigner's Friend

Tischler & her collaborators assumed that evaluating & conducting a friend of Wigner's experiment might put light on the quantum theory's limits. You're motivated by a recent wave of experimental & theoretical papers that have explored an observer's function in the quantum theory by taking entanglement in Wigner's traditional setup. Assuming you have two pieces of photons of light that are polarized such that they could still vibrate vertically or horizontally. The photons could also be put in a superposition by vibrating both vertically & horizontally at the same moment, like a paradoxical cat of Schrödinger can be both dead & alive until it is detected.

These photon pairs can be formulated together—entangled—so their polarizations are often observed to be in the reverse direction as observed. That does not sound strange—unless you note that these features are not set before testing. Even though one photon is allocated to a physicist named Alice in Australia, even when the other is shipped to her colleague Bob in the lab in Vienna, entanglement means that right after Alice notices her photon &, for example, discovers polarization for being horizontal, polarization for Bob's photon immediately synchronizes to vibrating vertically. Since two photons seem to interact quicker than the speed of light, something is forbidden by his theory of relativity. This effect greatly puzzled Albert Einstein, who called it "spooky behavior at a distance."

These problems remained speculative until the 1960s, although physicist "John Bell" invented a way to assess whether you're spooky or if there might be a more mundane cause for the associations between intertwined partners. Bell envisioned a commonsense idea that was local, one under which effects do not move among particles immediately. It was also deterministic instead of inherently probabilistic, meaning experimental outcomes might, in theory, be expected with confidence if only physicists knew enough about the system's unknown properties. It was practical, which, to the quantum physicist, implies that structures would have certain definite features even though nobody stared at them. Bell determined the highest degree of similarities between a set of intertwined particles that, like a deterministic, local, and practical theory, might sustain. If the threshold was breached during an experiment, all the hypotheses underlying the hypothesis must be incorrect.

These "Bell tests" have also been taken out, with the set of watertight iterations conducted in 2015, & they have verified the spookiness of reality. "Quantum principle is a discipline that established experimental results through Bell's [theorem]—now around 50 years old. And you have spent plenty of time re-implementing those tests & exploring whatever their mean," Steinberg claimed. "It is quite unusual that individuals are prepared to think of a new experiment that goes beyond Bell."

The Brisbane team's goal was to derive & evaluate some new theorem that will do just that, offering even tighter constraints— "local friendly nature" bounds—on the essence of truth. Like Bell's principle, the investigators' hypothetical one is local. These also expressly ban "super determinism," which demands that experimenters are responsible for choosing whether to calculate without being conditioned by incidents in the future or distant past. (Bell indirectly believed that experimenters would make free decisions, too.) Ultimately, the team recommends that anytime an observer performs a calculation, the result is a true, single occurrence in the world—it is not subjective to someone or something.

Checking local friendliness involves the cunning arrangement containing 2 "super observers," Alice & Bob (that play the part of Wigner), experiencing their friends Debbie & Charlie. Bob and Alice both have their very own interferometer—an instrument designed to control beams of photons. Until being Youighed, photons' polarizations are now in a superposition being both vertical & horizontal. Pairs of intertwined photons are formulated, so if each one's polarization is calculated to be the horizontal, polarization of its companion can automatically flip to be vertical. From every entangled pair, one photon is sent in the interferometer of Alice, as well as its companion is sent to the interferometer of Bob. Debbie & Charlie are not probably human companions in this exercise. Instead, they beam displacers at the front of every interferometer. As Alice's photon enters the displacer, polarization is efficiently determined, whether left or right, based on the polarization orientation this snaps through. This behavior performs the character of a friend of Alice Charlie "measuring" polarization. (Debbie similarly lives in the interferometer of Bob.)

Alice also needs to make some choices: She should automatically calculate the photon's current deviated direction, which will be equivalent to unlocking the lab door & telling Charlie what she had seen. Or she will enable photon to proceed on its way, going via a 2nd beam displacer which recombines left & right paths—equivalent of holding lab doors locked. Alice will then precisely calculate her polarization of photon when it leaves the interferometer. Throughout the experiment, Bob and Alice separately chose which measuring options to consider and instead compare results to quantify the similarities found across a set of intertwined pairs.

Tischler & her collaborators carried 90,000 runs of trial. As predicted, the associations breached Bell's initial bounds—& crucially, individuals also breached the current threshold of local friendliness. The team might even adjust the system to tune the degree of entanglement among photons downwards by placing one pair until it reached its interferometer, subtly perturbing the partners' optimal equilibrium. As investigators run experiments with this significant degree of entanglement, they reached a stage where the associations still breached Bell's requirement but not the local friendliness. This finding showed that two sets of bounds are not identical and that the current local-friendliness restrictions are greater, Tischler says. "If you breach them, you understand so much about actuality," she says. Namely, suppose the hypothesis suggests that "friends" should be viewed as quantum structures. In that case, you should either relinquish locality, admit that observations don't have a single conclusion that observers should consent on or encourage super determinism. - of these alternatives has profound—&, to certain physicists, notably distasteful—application.

4.12 Reconsidering Reality

"The paper is a significant philosophical examination," says Michele Reilly, fellow benefactor of Turing, a quantum-figuring organization situated in New York City, who was not associated with the work. She notes that physicists considering quantum establishments have frequently battled to concoct an achievable test to back up their enormous thoughts. "I'm excited to see an analysis behind philosophical investigations," Reilly says. Steinberg calls the investigation "amazingly exquisite" and acclaims the group for handling the secret of the onlooker's part in estimation head-on.

Even though it is nothing unexpected that quantum mechanics compels us to surrender a rational suspicion—physicists realized that from Bell—"the development here is that You are a narrowing in on which of those presumptions it is," says Wharton, who was likewise not a piece of the examination. He notes, advocates of most quantum understandings won't lose any rest. Enthusiasts of retrocausality, like himself, have just tried for some degree of reconciliation with super determinism: in their view, it isn't stunning that future estimations influence past outcomes. In the interim, QBists and many-universes disciples, quite a while in the past, tossed out the prerequisite that quantum mechanics recommend a solitary result that each spectator should concur on.

Furthermore, both Bohmian mechanics and unconstrained breakdown models as of now joyfully jettisoned territory because of Bell. Besides, breakdown models say that a genuine visible companion can't be controlled as a quantum framework in any case.

Vaidman, who was likewise not engaged with the new work, is less enthused by it, be that as it may, and censures the distinguishing proof of Wigner's companion with a photon. The techniques utilized in the paper "are crazy; the companion must be visible," he says. Savant of material science Tim Maudlin of New York University, who was not a piece of the investigation, concurs. "No one thinks a photon is a spectator, except if you are a panpsychism," he says. Since no physicist addresses whether a photon can be placed into superposition, Maudlin feels the analysis needs chomp. "It precludes something—simply something that no one at any point proposed," he says.

Tischler acknowledges the analysis. "You would prefer not to overclaim what You have done," she says. The key for future trials will scale up the size of the "companion," adds colleague Howard Wiseman, a physicist at Griffith University. The most emotional outcome, he says, would include utilizing human-made brainpower, epitomized on a quantum PC, as the companion. A few savants have pondered that such a machine could have humanlike encounters; a position is known as the solid AI speculation, Wiseman notes; however, no one yet knows whether that thought will end up being valid. Yet, on the off chance that the speculation holds, this quantum-based fake general knowledge (AGI) would be minute. So, according to the perspective of unconstrained breakdown models, it would not trigger breakdown because of its size. On the off chance that such a test was run, and the nearby kind disposition bound was not disregarded, that outcome would suggest that an AGI's cognizance can't be placed into superposition. Thus, that end would recommend that Wigner was correct that awareness causes a breakdown. "I don't figure I will live to see an analysis like this," Wiseman says. "Yet, that would be progressive."

Reilly, nonetheless, cautions that physicists trusting that future AGI will help them home in on the central portrayal of the truth are taking a mixed-up approach. "It's not incomprehensible to you that quantum PCs will be changing in perspective to get you in the AGI," she says. "At last, you need a hypothesis of everything to fabricate an AGI on a quantum PC, period, full stop."

That prerequisite may preclude more vainglorious plans. The group likewise recommends more humble middle tests, including AI frameworks as companions, which bids Steinberg. That approach is "intriguing and provocative," he says. "It's getting possible that bigger and bigger scope computational gadgets could, indeed, be estimated in a quantum way."

Quantum physicist Renato Renner, at Swiss Federal Institution of the Technology Zurich "ETH Zurich," creates a significantly more grounded guarantee: whether future analyses can be completed, he says, the new hypothesis reveals to us that quantum mechanics should be supplanted. In 2018 Renner and his partner Daniela Frauchiger, at that point at ETH Zurich, distributed a psychological study dependent on Wigner's companion and utilized it to infer another conundrum. Their arrangement varies from that of the Brisbane group, including four spectators whose estimations can get caught. Renner and Frauchiger determined that if the spectators apply quantum laws to each other, they can wind up deducing various outcomes in a similar examination.

"The new paper is another affirmation that You disapprove of the current quantum hypothesis," says Renner, who was not engaged with the work. He contends that none of the present quantum translations can escape the purported Frauchiger-Renner Catch 22 without defenders conceding they don't mind whether the quantum hypothesis gives reliable outcomes. QBists offer the most acceptable ways to get out, because from the beginning, they say that quantum hypothesis can't be utilized to deduce what different onlookers will quantify, Renner says. "It stresses me, however: If everything is only close to home to me, how might I say anything pertinent to you?" he adds. Renner is currently chipping away at another hypothesis that gives many numerical principles that would permit one onlooker to work out what another should find in a quantum exploration.

The individuals who emphatically accept their number one translation correctly see little incentive in Tischler's examination. "On the off chance that you think quantum mechanics is undesirable, and it needs supplanting, this is helpful because it discloses to you know imperatives,"

Vaidman says. "however, I disagree that this is the situation—numerous universes clarify everything."

Until further notice, physicists should keep on settling on a truce about which translation is ideal or if a new hypothesis is required. "That is the place where You left off in the mid-twentieth century—You're truly befuddled about this," Reilly says. "In any case, these examinations are the correct intention for thoroughly considering it."

Chapter 5: Einstein's Theory of Relativity

"Everything must be created as meek as probable, but no easier."

Without exception, the peculiar relativity theory is one of the most influential developments during the history of science, and 2nd only to the observation by Newton of mechanics laws in their significance to physics. Despite this, peculiar relativity remains little known & there is a lot of confusion regarding the topic on the internet & in newspapers. Largely undeserved notoriety for being too hard for certain persons to grasp does not support this.

The simple concepts aren't that hard to grasp. By following a clear route through the evolution of physics from Galileo, this essay would clarify some of those fundamental theories, explaining why laws of physics since they were known in the 19th century would have to be modified, demonstrating how specific relativity emerged from that change, and discussing some of the implications of that latest theory.

5.1 Covariance, Reference frames & Galilean relativity

The fundamental principle of relativity is that physics rules should be agreed upon by two separate individuals who are really in motion compared to one another. When two independent observers are in a redshift, they are in separate reference frames, & such reference frames are assumed to be inertial when their relative velocity is unchanged. If all researchers accept a physical principle throughout the inertial reference frames, the theory is covariant. Here it would be best if you considered inertial frames only.

Assume that if an observer at rest concerning frame S is at the center of the coordinate system "x,y,z" and that an observer at rest concerning frame S is at origin of the coordinate system "x,y,z." If the S observer sees the center of S coordinates through constant velocity V shifting to right, so two reference frames seem to be in normal configuration:

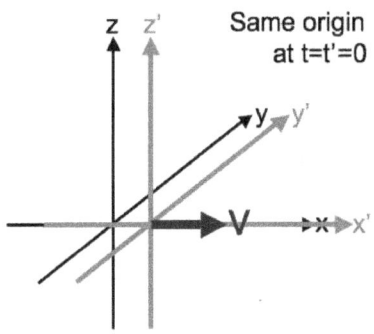

S & S' are still well-thought-out to be in a regular setting.

Assume the observer in S states that such an event happens at point P at the time T0 & that another incident occurs at time T1 at point Q and that L is the gap between P & Q and $\Delta T=T_1-T_0$. Assume that an individual in S' sees the similar events, distinguished by interval L' and with the 2nd occurrence T' seconds after first. The following conclusions You're made before Einstein:

Distance is absolute: L=L'

Time is absolute: $\Delta T=\Delta T'$

Lengths are specified by Pythagorean theorem: $L^2=(\Delta x)^2+(\Delta y)^2+(\Delta z)^2$ where Δx, Δy, and Δz are The displacements in the directions of x, y, and z. Quantities that in all inertial frames have similar numerical values are considered to be unobservable.

Let x, y, z, & t be coordinate system's location & time coordinates attached to S and x', y', z', and t' The coordinate system's position & time coordinates attached to S'. These distance and time interval assumptions mean that the following law interconnects these coordination mechanisms.:

$$x' = x - Vt$$
$$y' = y$$
$$z' = z$$
$$t' = t$$

This is considered the transformation of the Galileans. The idea that the rules of physics science are covariant in comparison to Galilean transformation is Galilean relativity. For physics of Newton, you will verify if this is valid. Suppose the rule of Newton is valid in frame S'

$$m\frac{d^2 x'}{dt'^2} = F'(x', t')$$

the y' and z' coordinates are being ignored for simplicity.

Here, F'(x',t') is an experimentally determined arbitrary color, & F(x,t) according to the frame S is a term for same force. Suppose that, by testing quantitative value for x' & t', the investigator in S' calculated F'(x',t'). The observer in S tests the same quantitative value for x & t to locate F(x,t). Because it is the same color, & since S has no acceleration than S, the investigator in S would obtain the same effects, F(x,t)=F'(x,t '). You get through plugging x'=x-Vt & t'=t in the derivative:

$$\frac{d^2x'}{dt'^2} = \frac{d^2x'}{dt^2} = \frac{d^2}{dt^2}(x - vt) = \frac{d^2x}{dt^2}$$

It follows that:

$$m\frac{d^2x}{dt^2} = F(x, t)$$

So, physics of Newtonian is covariant to transformation of the Galilean. But does this extend to all the rules of physics?

Recognize the equations of Maxwell in a space region free of currents or charges:

$$\nabla \cdot \mathbf{E} = 0$$
$$\nabla \cdot \mathbf{B} = 0$$
$$\nabla \times \mathbf{E} = -\frac{\partial \mathbf{B}}{\partial t}$$
$$\nabla \times \mathbf{B} = \frac{1}{c^2}\frac{\partial \mathbf{E}}{\partial t}$$

Using curl of third line & identity of vector calculus $\nabla \times (\nabla \times E) = \nabla(\nabla \cdot E) - \nabla^2 E = -\nabla^2 E$ since $\nabla \cdot E = 0$. The variable E of an electric field then obeys wave equation:

$$\nabla \times (\nabla \times \mathbf{E}) + \nabla^2 \mathbf{E} = 0$$
$$\therefore \nabla \times \left(-\frac{\partial \mathbf{B}}{\partial t}\right) + \nabla^2 \mathbf{E} = 0$$
$$\therefore -\frac{\partial}{\partial t}(\nabla \times \mathbf{B}) + \nabla^2 \mathbf{E} = 0$$
$$\therefore \nabla^2 \mathbf{E} - \frac{1}{c^2}\frac{\partial^2 \mathbf{E}}{\partial t^2} = 0$$

By the same phase, vector B or magnetic field can also follow wave equation. This equation assumes that the speed of light can transmit a disruption in the electric field with some constant velocity c. Imagine that electric field, the only other part of which is in the direction of x and doesn't rely on y or z. Assume wave equation in frame S' is obeyed:

$$\frac{\partial^2 E'}{\partial x'^2} - \frac{1}{c^2}\frac{\partial^2 E'}{\partial t'^2} = 0$$

you require this to transform in:

$$\frac{\partial^2 E}{\partial x^2} - \frac{1}{c^2}\frac{\partial^2 E}{\partial t^2} = 0$$

Observe if this occurs with Galilean transform.

The results transform corresponding to chain rule:

$$\frac{\partial}{\partial x'} = \frac{\partial x}{\partial x'}\frac{\partial}{\partial x} + \frac{\partial t}{\partial x'}\frac{\partial}{\partial t} = \frac{\partial}{\partial x}$$

$$\therefore \frac{\partial^2}{\partial x'^2} = \frac{\partial^2}{\partial x^2}$$

$$\frac{\partial}{\partial t'} = \frac{\partial x}{\partial t'}\frac{\partial}{\partial x} + \frac{\partial t}{\partial t'}\frac{\partial}{\partial t}$$

$$= V\frac{\partial}{\partial x} + \frac{\partial}{\partial t}$$

$$\therefore \frac{\partial^2}{\partial t'^2} = \frac{\partial}{\partial t'}\left(V\frac{\partial}{\partial x} + \frac{\partial}{\partial t}\right)$$

$$= \frac{\partial x}{\partial t}\frac{\partial}{\partial x}\left(V\frac{\partial}{\partial x} + \frac{\partial}{\partial t}\right) + \frac{\partial t}{\partial t'}\frac{\partial}{\partial t}\left(V\frac{\partial}{\partial x} + \frac{\partial}{\partial t}\right)$$

$$= V^2\frac{\partial^2}{\partial x^2} + 2V\frac{\partial^2}{\partial x \partial t} + \frac{\partial^2}{\partial t^2}$$

And this suggests that the wave equation as shown in frame S' under Galilean transformation turns into the following as observe from S:

$$\left(1 - \frac{V^2}{c^2}\right)\frac{\partial^2 E}{\partial x^2} - \frac{2V}{c^2}\frac{\partial^2 E}{\partial x \partial t} - \frac{1}{c^2}\frac{\partial^2 E}{\partial t^2} = 0$$

This poses a problem: spectators will conflict with the rule regulating the transmission of a ray of light in distinct inertial frames. You have no alternative but to assume that certainly one of the resulting propositions is valid to address this.:

- The calculations of Maxwell are incorrect.

- There is just one unique reference point, the residual frame of so-called aluminiferous, in which equations of Maxwell are valid.

- The Galilean transformation is false, and so are fundamental concepts about space & time.

You should automatically discard the first proposition. Experimental truth is equations of Maxwell. The 2nd should be dismissed considering the many decades in which scientists sought and struggled to locate in the latter half of the 19th century. This leaves the third choice.

5.2 The Lorentz Transform & theory of relativity (Einstein)

In 1892, a paper was published by Hendrik Lorentz in which he demonstrated that the transition under which equations of Maxwell are covariant is:

$$x' = \gamma(x - Vt)$$
$$y' = y$$
$$z' = z$$
$$t' = \gamma\left(t - \frac{Vx}{c^2}\right)$$

Where γ is known as Lorentz factor:

$$\gamma = \frac{1}{\sqrt{1-\frac{V^2}{c^2}}}$$

With the necessary change, the laws of Newton are indeed covariant under such a transition.

This is known as Lorentz transform. Unfortunately, Lorentz could not provide a proper physical explanation, as the Planet's motion compared to the luminous was mistakenly credited to him.

In his book in 1905 On Electrodynamics of surrounding Bodies, Einstein presented the right explanation, & this interpretation is the basis of what was considered special relativity. With the following 2 postulates, he began:

- The rules of physics are the same in all the inertial reference points.

- The light's speed has the same meaning in all the inertial reference points, which is, it's invariant.

You will utilize this to extract Lorentz transition, although this would entail several improvements to how space & time are interpreted.

5.3 Time Dilation

Let S' be a remaining frame of the train with S in the normal setup, the remaining frame of everyone on the platform. On a train, the experiment is conducted in which any physical procedure takes place over the time Δt'. You would illustrate that platform investigator will find that the similar physical procedure is happening over the period between Δt, while Δt & Δt' are connected by:

$$\Delta t = \gamma \Delta t'$$

Because γ>1, this is known as time dilation. Assume that the laser pulse exits point A, moves straight upwards & then bounces from the mirror at point B, according to the individual experimenting on the train, & then returned to the detector at point C. That is straight next to A'.

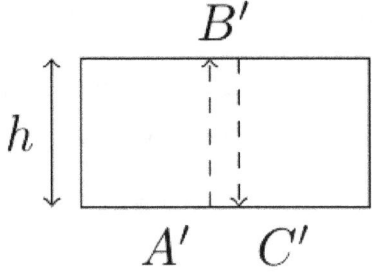

The average distance traveled by laser pulse was two hours & speed was c, therefore:

$$\Delta t' = \frac{2h}{c}$$

Let's now thought about what analyst sees on site. As laser pulse passes from emitter to mirror & then back to a detector at steady rpm, the train still shifts to the right. Laser pulse direction is the triangle for an observer on the platform:

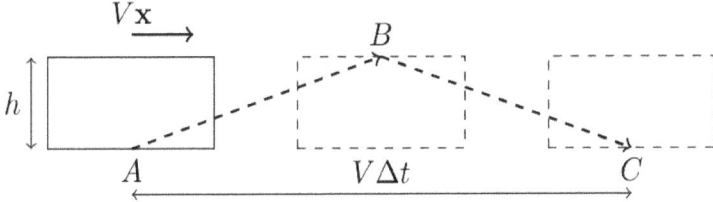

From Pythagorean Theorem, length of line AB is presented by:

$$|AB| = \sqrt{h^2 + \left(\frac{V\Delta t}{2}\right)^2}$$

The full path length is double this number, and because the

speed of light in all reference frames is the same, overall path length should be equivalent to cΔt: hence

$$\Delta t = \frac{2}{c}\sqrt{h^2 + \left(\frac{V\Delta t}{2}\right)^2}$$

You are now solving the Δtin form of Δt' by removing h. You have h=cΔt'/2 from formula for Δt'; then you can get h=cΔt'/2 by connecting this in & squaring all sides of a formula for Δt:

$$(\Delta t)^2 = \frac{4}{c^2}\left(\frac{c^2}{4}(\Delta t')^2 + \left(\frac{V\Delta t}{2}\right)^2\right)$$

$$= (\Delta t')^2 + \left(\frac{V}{c}\right)^2 (\Delta t)^2$$

$$\therefore \left(1 - \left(\frac{V}{c}\right)^2\right)(\Delta t)^2 = (\Delta t')^2$$

$$\therefore \Delta t = \frac{\Delta t'}{\sqrt{1 - \left(\frac{V}{c}\right)^2}}$$

$$= \gamma \Delta t'$$

This illustrates that time is constipated among reference points since the speed of light should remain invariant.

This describes why running clocks tend to operate relatively slow: if the clock on the train ticks each second, according to those observing from tracks, the gap amongst ticks is increased.

5.4 Demo: Muon decay

A muon is indeed a subatomic particle which, except for its density, is in any way like either electron: muon is around heavier 207 times. Interaction allows the electron & two additional particles known as electron antineutrino & muon neutrino to degrade in the muon.:

$$\mu \rightarrow e + \bar{\nu}_e + \nu_\mu$$

Muons do have a decaying half-life of approximately 2.2 microseconds, indicating that it would take around 2.2 microseconds with half of them just to degrade in the electrons if you've got a specimen of 100 muons.

Muons are formed at an altitude of around 15 km in the upper atmosphere whenever cosmic rays hit gas molecules. Muon detectors usually detect 1 muon/square centimeter/minute at sea level, & their average velocity is around 0.995c once they're observed at sea level. You will find the time that it takes for one muon to hit sea level is around 15,000m/0.995c~50 microseconds, or around

23 half-lives, if you neglect relativity. Because the flux of the muons at the sea level is 1/s·cm2, this implies that 223/s·cm2 will be several muons produced at the altitude, which is not a very reasonable number.

You will have to see what occurs whenever special relativity is considered. It does not matter what more time passed for you, the spectator, whenever it refers to the particle decay. What counts is how long time passes for particles; however, if such particles travel much quicker than you, then a shorter period $\Delta t'=\Delta t/\gamma$ elapses for particles due to time dilation.

In the case of V=0.995c, γ~10

So, while it appears to you that the muons require about 50 microseconds to hit sea level, it requires even almost 5 microseconds, or around 2.3 half-lives, for muons. This suggests that if muon flux at the sea level is 1/s·cm2, then muon the flux at 15 kilometers is generated at around 4.5/s·cm2, which is a far more rational rate.

For their side, the muons see the earth rising at a velocity of 0.995c towards them. After just 5 microseconds, muons could see themselves touching the earth; how else could they have flown 15,000 meters within 5 microseconds at the pace of just 0.995c? The response is that You're not.

5.5 Length Contraction

In frame S, an investigator at rest sees a particle with the velocity Vx pass the post at level A at time t=0, & at time t, she saw particle pass the post at point B, divided by length L on both the points on the x-axis. The particle's remaining frame is S', & the particle is stationary in S' & two posts, divided by duration L', approach particle with the velocity -Vx. At time t'=0, the first post passed particle & at time t'= t'/γ second post passes particle. You will observe that Δt'=Δt/γ. Since L=VΔt and L'=VΔt'. This suggests that in the moving picture, the length is contracted.

This addresses the issue asked at the end of the last section: muons do not have to traverse 15,000 meters measured on the ground by the observer. The distance in the rest frame of the muon is just 1500 meters.

According to the observer, this corresponds to the reduction of space in the residual frame of a moving entity. There is also a reciprocal variant of this theory, which notes that it appears to contract moving objects. Let S be the remaining frame of the rod that seems to be rotating according to observatory frame S with velocity

Vx. There is no way for the spectator to understand that shifting proportional to rod or rod is shifting compared to her. If she shifted relative to the rod, she'd see the room as being compressed such that the rod seems to be shorter than the duration of its rest. This is precisely equal to suggesting that the rod seems smaller

when it runs, since she cannot tell if she or the rod is shifting.

5.6 Lorentz Transform

You may now show that Lorentz transform connects coordinate structures of two reference frames in a normal setup. You're going to explain that:

$$x' = \gamma(x - Vt)$$
$$y' = y$$
$$z' = z$$
$$t' = \gamma\left(t - \frac{Vx}{c^2}\right)$$

Because the velocity of S' proportional to S is fixed & in x-direction completely, it must be valid by symmetry that y'= y & z'= z.

You are now describing the new quantity named spacetime interval (Δs)² to continue.:

$$(\Delta s)^2 = c^2(\Delta t)^2 - (\Delta x)^2 - (\Delta y)^2 - (\Delta z)^2$$
$$(\Delta s')^2 = c^2(\Delta t')^2 - (\Delta x')^2 - (\Delta y')^2 - (\Delta z')^2$$

For all the pairs of the reference frames S & S ', you can demonstrate that the spacetime period is invariant, indicating that (Δs)²=(Δs')².

Assume that man standing still very much in the frame S throws out a laser pulse which travels L distance in the time Δt', & pulse travels L distance in time Δt' concerning frame S. Then c2=(L/t)2=(L'/t')2 through c's invariance. You've had then:

$$c^2(\Delta t)^2 - L^2 = 0 = c^2(\Delta t')^2 - (L')^2$$
$$\therefore c^2(\Delta t)^2 - (\Delta x)^2 - (\Delta y)^2 - (\Delta z)^2 = c^2(\Delta t')^2 - (\Delta x')^2 - (\Delta y')^2 - (\Delta z')^2$$
$$\therefore (\Delta s)^2 = (\Delta s')^2$$

The remainder of Lorentz transform distribution fits derivation used by Einstein in his influential book on relativity. In the second line above, remove delta symbols & when y'=y & z'=z, these words null out of the equation. You will compose then:

$$c^2t^2 - x^2 = c^2(t')^2 - (x')^2$$
$$\therefore (ct-x)(ct+x) = (ct'-x')(ct'+x')$$

You can utilize this to write:

$$x' - ct' = \lambda(x - ct)$$
$$x' + ct' = \mu(x + ct)$$

To get the equation for x' & deduct the first equation from the second, apply the second equation to the first one to get the equation for ct':

$$2x' = (\mu + \lambda)x + (\mu - \lambda)ct$$
$$2ct' = (\mu - \lambda)x + (\mu + \lambda)ct$$

Then make up the following tasks:

$$a = \frac{\mu + \lambda}{2}$$

$$b = \frac{\mu - \lambda}{2}$$

So, you will get the linear system for x' & Ct':

$$x' = ax + bct$$

$$ct' = bx + act$$

There is a velocity Vx at the root of the primed coordinates scheme so that you can set its location vector with (Vt,0,0), so let x'=0 fit x=Vt. The 1st equation then provides:

$$0 = aVt + bct \implies b = -\frac{av}{c}$$

Now the system of these equations come to be:

$$x' = a(x - Vt)$$

$$ct' = -ac\left(t - \frac{Vx}{c^2}\right)$$

Plug this into the term for invariance of space period to overcome for a, c²(t')²-(x')²=c²t²-x²:

$$(ct')^2 - (x')^2 = a^2c^2\left(t - \frac{Vx}{c^2}\right)^2 - a^2(x - Vt)^2$$

$$= a^2\left(c^2\left(t^2 + \frac{V^2}{c^4}x^2 - \frac{2Vxt}{c^2}\right) - x^2 - V^2t^2 + 2Vxt\right)$$

$$= a^2\left(c^2t^2 + \frac{V^2}{c^2}x^2 - 2Vxt - x^2 - V^2t^2 + 2Vxt\right)$$

$$= a^2\left((c^2 - V^2)t^2 - \left(1 - \frac{V^2}{c^2}\right)x^2\right)$$

$$= a^2\left(\left(1 - \frac{V^2}{c^2}\right)(ct)^2 - \left(1 - \frac{V^2}{c^2}\right)x^2\right)$$

$$= a^2\left(1 - \frac{V^2}{c^2}\right)\left((ct)^2 - x^2\right)$$

$$= (ct)^2 - x^2$$

This implies that:

$$a^2 \left(1 - \frac{V^2}{c^2}\right) = 1$$

$$\therefore a = \frac{1}{\sqrt{1 - \frac{V^2}{c^2}}} = \gamma$$

So, when You replace this with formulas you will found for x & ct, you will get Lorentz transform:

$$x' = \gamma(x - Vt)$$
$$y' = y$$
$$z' = z$$
$$t' = \gamma\left(t - \frac{Vx}{c^2}\right)$$

So, you have effectively extracted Lorentz from physical concepts to transform.

5.7 Demo: traditional limit

The Lorentz transformation appears very distinct from the transformation of the Galilean t. How could physicists have been so wrong for so long in this regard? Consider a fighter aircraft flying with V=350m/s above the sound velocity compared to an observer present in a neighboring control. Then $V^2/c^2 \sim 1.4 \times 10^{-12}$. For small values of V2/c2, the easiest approach to estimate the value of γ is to use the binomial approximation, which specifies that:

$$(1+x)^\alpha \approx 1 + \alpha x \text{ when } |x| << 1$$

It provides a reasonable estimation for the value of γ:

$$\left(1 - \frac{V^2}{c^2}\right)^{-1/2} \approx 1 + \frac{V^2}{2c^2} \approx 1 + 7 \times 10^{-13}$$
$$= 1.0000000000007$$

Therefore, non-relativistic physics is also precise to parts per trillionth for speed like sound speed (12 decimal places). And of course, until the 20th century, even this "low" velocity

was virtually inaccessible to those doing research. It would probably never have been observed by someone in their everyday lives, so it's easy to understand why it took about 300 years after Galileo's period before anyone realized anything was incorrect.

5.8 Spacetime

It cannot be highlighted enough that time dilation and contraction of length are features of space and time itself. Besides this, traveling at relativistic speed also cannot produce forces that expand or compress artifacts. Therefore, it is not the result of an optical illusion and measuring error that allows any object's length or the amount that a clock ticks to be misjudged by observers in various frames. When observers record various lengths for measuring rods or different frequencies for ticking clocks in separate frames, they all are considered right since time intervals and length measurements are not invariant. Space and time function exactly that way.

Classical physics is composed of three-dimensional Euclidean space, E3, the group of all arranged triples of real numbers (x,y,z) coupled with an adequate topological structure to make aspects like "distance" and "point"

significant, as a feature called the Euclidean metric. Euclidean metric states that the distance among two points P1=(x1,y1,z1) and P2=(x2,y2,z2) is:

$$d(P_1, P_2) = \sqrt{(x_2 - x_1)^2 + (y_2 - y_1)^2 + (z_2 - z_1)^2}$$

In classical physics when an event occurs at time t1 at point p1. Then another event occurs at time t2 at P2. Here, P1 and P2 are separated by t (time) and L (distance)= t2-t1. It can be explained as the two events that occurred so that they are separated by L meter. The second event took place after the first with an interval of Δt seconds. If You claim that space and time are "separate" in classical physics, this is what it means. There is no logical way of assigning a single number as the "distance" in classical spacetime between two cases.

Presently, do You live in Euclidean space?

No, You don't. If spacetime was considered Euclidean. Galilean transformation would be considered the correct relationship among the coordinates of multiple reference frames, but distances would be invariant concerning a changed reference frame. But because of the contraction in length, this is false. This raises the question about what kind of space is in which You are currently living.

Considering the group of all spacetime points (x,y,z,t), assume this time when the spacetime interval is the "distance" between two points. If event s1 occurs at position (x1,y1,z1) and time t1, and another event s2 occurs at position (x2,y2,z2) at time t2, the "distance" among the following events is given by:

$$d(s_1, s_2) = (\Delta s)^2 = c^2(t_2 - t_1)^2 - (x_2 - x_1)^2 - (y_2 - y_1)^2 - (z_2 - z_1)^2$$

After Herman Minkowski, the title for the function that gives the spacetime interval among two events is the Minkowski metric. The one who formalized the notion of spacetime was Minkowski. He was one of Einstein's college professors. Instead of a three-dimensional Euclidean space with one "extra" time dimension, you live in four-dimensional Minkowski spacetime. The consequences of this are vast, and several will have to wait for this article to be followed up. But let's discuss the most famous one that is close to this article.

Demo: equivalence of Mass-energy, $E=mc^2$

Among the most famous significance of specific relativity is that energy is equivalent to rest mass. A particle's rest mass is its mass determined in the frame in which the particle does not move. This portion is intended to provide justified

proof but not a formal one for this argument.

Firstly, through physical principles, it would be argued that if a particle has no rest mass, it can move with light speed. The real facts will have to wait for this article to be followed up. which would be about certain relativistic physics's implementations

Suppose a particle appears to travel in frame S at light speed. It travels a distance L in time Δt so that $c\Delta t = L$, so $(\Delta s)^2 = (c\Delta t)^2 - L^2 = 0$. however, by the invariance of intervals in space-time, $(s')2=(s)2$, so in any other reference frame S', $(s')2=(c't')2-(L')2=0$ so L'/t'=c, so in every inertial frame, the speed of the particle is c. This implies that it lacks a rest frame, so it is not physically important to say that it has a rest mass. It explains the "only if" portion of the statement.

Now presume a frame that contains a particle at rest and has zero mass; thus, the particle may not exist either. As it has no momentum to transfer to other particles, it is at rest. It has no mass; therefore, it is impossible to receive momentum from other particles too. Consequently, that particle cannot communicate with anything in this universe. You may assume that there is no frame in which a massless particle is at rest because You are only concerned with particles that show meaningful

physical existence. Therefore, all massless particles must move at light speed. This explains the "if" portion of the argument.

In the sequel, you can find that relativity leads to the different working of momentum and energy than how people usually speak of them. However, the fundamental conservation laws still hold in each reference frame; energy and momentum are still conserved.

A positron is a subatomic particle and antiparticle to the electron. In any sense, it is like an electron, with the distinction of having an opposite charge. It is understood from experimentation that they annihilate each other and create radiation when a particle and an antiparticle meet each other. For this, the formula is:

$$e^+ + e^- \rightarrow 2\gamma$$

E- represents an electron, e+ represents a positron, and γ means a photon; it generates two photons. Consider the situation just before annihilation happens; both the electron and positron are at rest and in touch with each other. In the system, the total mass is 2me, double the mass of an atom. Zero is the total momentum in the system. The

total rest mass is zero after the demolition. As a result, photons move with light speed. The question arises, where does all mass go, and from where energy comes?

Although momentum remains conserved, the overall momentum is still zero even after destruction. So both photons possess opposite directions but the momentum of the same p magnitude. Since there is no resting mass of photons, they do have kinetic energy, which You may write as $E=pc$.

Experiments also show that two photons' total kinetic energy is around 1.637-10-13 Joules after the destruction. Two electrons have a gross rest mass of around 1.829-10-30 kilograms. When You multiply this total rest mass by c2, you get $(2m_e) C^2 = 1.644 \times 10^{-13}$ Joules, which is the same as the total energy emitted. During a rounding error. As this experiment is performed again with neutrons and antineutrons, antiprotons and proton, muons, and anti-muons, and many more, the resulting relationship between the rest mass and energy remains the same.

As energy remains conserved, the system's energy level must have been the same before and after the demolition. This shows that it could be hypothesized that the energy was retained as the mass of the positron and

electron before the demolition, with the quantity of energy stored being $E=mc2$. This energy was transformed into the kinetic energy of the photons by the annihilation method. This will account for the presence of kinetic energy in a system that initially had no kinetic energy and eliminated the rest mass in the system; however, this must be proved later.

-Disclaimer-

All images that are without any reference belong to my original work. Some examples explained here are based on examples provided in "MODERN PHYSICS FOR SCIENTISTS AND ENGINEERS" AND FROM ITS SECOND EDITION WRITTEN BY Dubson, Zafiratos and Taylor.

I hope you're enjoying the reading, if YES, please do not forget to leave me a short review

Thanks for spending your time

Chapter 6: Simple Made Entanglement

One of the trickiest ideas in science is Quantum entanglement, but the key problems are basic. But once learned, entanglement sets up a deeper awareness of ideas such as quantum theory's "many worlds."

James O'Brien for Quanta Magazine

An atmosphere of glamorous mystery relates to the idea of quantum entanglement, as well, as some related claims show that "many worlds" are required by quantum theory. But these are, or should be, scientific theories in the end, with implications and some realistic meanings. Here, the concepts of entanglement and many worlds will be clarified simply as possible.

I.Entanglement is sometimes perceived as a unique quantum-mechanical phenomenon; however, it is not. It is a fact that it is unconventional though enlightening; firstly, it is considered as a simple non-quantum or "classical" version of entanglement. This allows us to inquire about the sensitivity of entanglement separately from the common oddity of quantum theory.

6.1 Quantized

A monthly column in which presents the work of top researchers exploring the discovery process. Frank Wilczek is the columnist of this month who is a Nobel Prize-winning physicist working at the Massachusetts Institute of Technology.

Entanglement occurs in cases where You have limited information about the condition of two processes. For example, two objects that You'll call c-ons may form your systems. The "c" is supposed to suggest "classical," so you might think of your c-ons as cakes if you'd like to have something unique and interesting in mind.

Your c-ons are available in two shapes which You define as their potential states. Then, for two c-ons, the four feasible joint states are (circle, circle) (square, square), (square, circle), and (circle, square). In each of those four states, the following tables show two instances of the possibilities for

locating the system in all states.

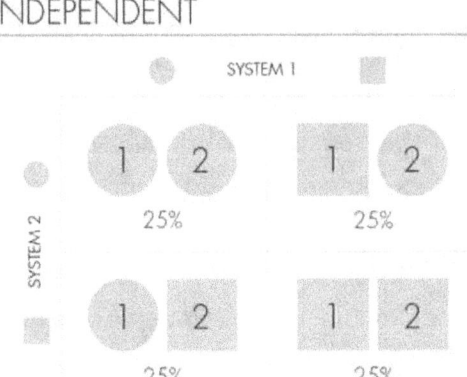

Quanta Magazine/Olena Shmahalo

"If information about the state of one of them does not provide useful data on the state of the other, you claim that the c-ons are independent. This property is explained in first table. If first c-ons (or cake) are square, so You still don't know any information regarding the second's form. Likewise, the second's form would not disclose something useful about the first's shape.

On the other side, as details regarding one strengthen your perception of the other, you claim your two c-ons are intertwined. Extreme entanglement is illustrated in your second table. In that case, you realize the second cake is circular, too, whenever the first c-ons are circular. And whenever square is the first c-on, so is the second too. When you know the structure of one, you may confidently assume the type of the other.

The quantum form of entanglement is the same phenomenon that lacks independence. In quantum theory, mathematical objects are used to describe states known as wave functions. Very interesting complications are introduced by the instructions connecting wave functions to physical probabilities. But the core principle of intertwined information, as for classical probabilities You have already discussed, carries over.

Of course, cakes cannot be counted as quantum systems. Still, entanglement among quantum systems occurs normally in the aftermath of particle collisions.

One practical work, unentangled (independent) states, is an

unusual exception, as interaction generates associations between them whenever structures communicate.

They are considering molecules as an example. Subsystems include composites, including electrons and nuclei, the lost energy condition of a molecule, in which it is most frequently observed, is the strongly entangled state of its nuclei and electrons. since the locations of certain constituent particles cannot be isolated in anyways. The electrons shift with them while the nuclei move.

Explaining the above examples: explaining square or circular states of system 1, it is written as $\Phi_\blacksquare$, $\Phi_\bullet$ for the wave functions. While for system 2, it is written as $\psi_\blacksquare$, $\psi_\bullet$ for the same states. In this working example, the whole states will be:

Entangled: $\Phi_\blacksquare \psi_\blacksquare + \Phi_\bullet \psi_\bullet$

Independent: $\Phi_\blacksquare \psi_\blacksquare + \Phi_\blacksquare \psi_\bullet + \Phi_\bullet \psi_\blacksquare + \Phi_\bullet \psi$

Independent version can be written as:

$$(\Phi_\blacksquare + \Phi_\bullet)(\psi_\blacksquare + \psi_\bullet)$$

Note: in what way in this design, the parentheses distinct systems 1 & 2 in the independent parts.

There are several ways that entangled states can be formed. One approach is to calculate the composite system that is providing partial results. You will be explained through example that two structures have conspired to have the same structure without knowing precisely what form they have. Later, this notion would become important.

The more distinctive implications of the quantum entanglement, such as the Greenberger Horne Zeilinger (GHZ) impacts and Einstein-Podolsky-Rosen (EPR), emerge from its relationship with another component of the quantum theory named "complementarity." Complementarity is required to be introduced for discussion of EPR and GHz.

Previously, you figured that your C-ons should represent two shapes (square and circle). You now imagine the two colors, red and blue, will also be exhibited along with shapes. Talking about classical structures, this added property would mean that your c-ons might be in either of four potential states. If You're talking about classical structures, possible results will include a red circle, red square, blue circle, or blue square.

However, with a quantum cake, maybe a quake, or a q-on (with more dignity), the case is radically different. A q-on will show various shapes of different colors in different conditions does not imply that it simultaneously has both a form and a color. As You'll see soon, the "common sense" inference that Einstein believed should be part of every reasonable definition of physical reality is inconsistent with experimental evidence.

The shape of q-on can be measured, but it will lead to the loss of all information. Similarly, the color of q-on can also be measured, but it will be losing all information about its shape of q-on. According to quantum theory, it is not possible to measure both its color and shape simultaneously. One view of physical reality never explains all its aspects at once. As Niels Bohr framed it, it is the core of complementarity.

As a result, in assigning observable existence to individual properties, quantum theory forces one to be circumspect. It would help if you accepted the followings to prevent contradictions

> **1.** A property that cannot be measured needs no existence.
>
> **2.** Measurement is a vigorous process that changes the system that is being measured.

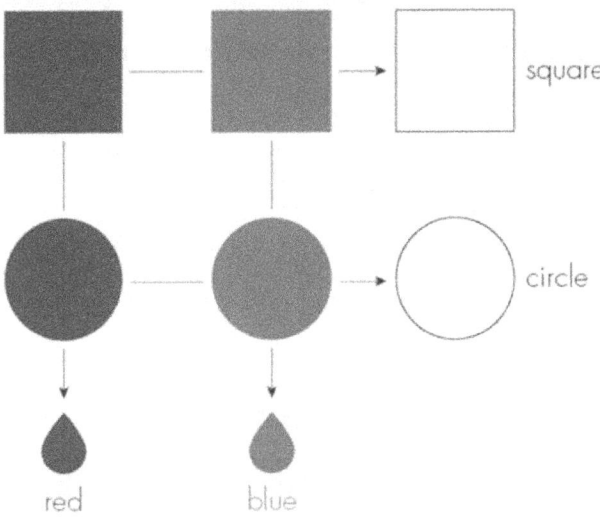

II.

Two classics will be discussed here now, which are far from classical and illustrations of quantum theory's strangeness. In rigorous tests, these have been confirmed. (In real experiments, people calculate properties such as electron's angular momentum rather than cakes' shapes or colors.)

A surprising effect of two quantum systems is Albert Einstein, Nathan Rosen (EPR), and Boris Podolsky. The EPR effect joins with an experimentally realizable type of quantum entanglement along with complementarity.

There are two q-ons in an EPR pair, one of which can be calculated either for its form or color (but not possible for both). You presume that several such pairs are all similar and that You may select which measurements to create from their components. You notice it is equally probable to be square or circular if You calculate the form of one element of an EPR pair. You notice it is equally probable to be red or blue whether You calculate the color.

The fascinating effects, which EPR proved paradoxical, originates when Your measurements of all pair members are taken. When You measure both members for shape and colors, you discover that the consequences always agree. Consequently, if You find that one is red and later check the other's color, you will discover that it will be red too, and so forth. Besides this, if You measure the shape of one and later the other's color, there will be no correlation. Therefore, if the first comes in the square, then the second is correspondingly red or blue.

According to quantum theory, even though great distances divide the two systems, you can get certain effects, and the calculations are done almost simultaneously. In one location, the selection of measurement tends to influence the system's status in the other location. As Einstein named it, this "spooky action at a distance" may appear to entail the transfer of knowledge at a rate greater than the speed of light, in this case, details on the calculation was done.

But is it? You don't know what to think until you obtain the results. Only when You learn the effect you've calculated, valuable knowledge is obtained, not when it is calculated. The result revealing message needs to be transmitted in concrete physical ways, presumably slow compared to light speed.

The paradox disintegrates further upon further thought. Let us consider the second system's condition again, as the first was calculated to be red. If You want to test the color of the second q-on, you'll get red. But when adding complementarity, as You mentioned earlier, if You want to calculate the form of a q-on, when it is in the "red" condition, you will have an equal chance of having a square or a circle. Therefore, the EPR result is theoretically induced, far from implementing a paradox. It is a repackaging of complementarity.

Neither is it paradoxical to find that there is a connection between distant events. And besides, if a pair of gloves are placed in boxes for each member and mail them to opposite directions of the world, one should not be shocked that one will evaluate the glove's handiness in the other from looking inside one package. Similarly, when the participants are close together, the correlations of an EPR pair must be imprinted. Although in all established circumstances, they will withstand subsequent separation as they had memories. Again, there is no connection with a peculiarity of EPR as such, but its potential expression is complementary.

III.

Daniel Greenberger, Anton Zeilinger, and Michael Horne found another beautifully illuminating example of quantum entanglement. Three of your q-ons are used, prepared in a unique, entangled condition (GHZ state). The three q-ons are being spread to three remote experimenters. Each experimenter selects whether to calculate form or color, individually or at random and notes the outcome. The procedure is replicated several times, always beginning with the GHZ Staten case of all three q-ons.

Each experimenter finds sufficiently random outcomes independently. She is equally capable of finding a square or a circle when she determines a q-on's shape. Similarly, when she calculates its color, red or blue are equally likely. So far, so earthly.

But later, a bit of study shows a stunning finding as the

experimenters step together and discuss their measurements. Let's use the label "good" for square shapes and red colors and "evil" for circle shapes and blue colors. Experimenters noticed that when two of them wanted to measure shape but the third measured color, they realized that "evil" was exactly 0 or 2 effects (circular or blue). But they learned that precisely 1 or 3 measures You're evil as all three decided to quantify hue. It is what is expected by quantum mechanics, and that is what is observed.

So: is the evil accounts even or odd? Both possibilities are realized in various kinds of measurements with confidence. You're obliged to ignore the issue. Speaking of the sum of bad in your system, regardless of how it is calculated, makes little sense. Sometimes, it contributes to inconsistencies.

The GHZ effect is explained by "quantum mechanics in your face," in the scientist Sidney Coleman's words. It demolishes a profoundly established belief entrenched in everyday practice. It explains that physical systems have definite properties, regardless of whether such properties are calculated. For if they did, then measurement options would not impact the equilibrium of good and evil. The message of the GHZ influence is unforgettable and mind-blowing, once internalized.

IV.

So far, you have discussed how entanglement will make it difficult to assign many q-ons to a delegate's unique, autonomous state. Similar contributions can be applied to a single q-on in time for its evolution.

You claim that You have "entangled histories" because it is difficult to attribute a definitive status to your framework at any point in time. You may produce entangled history by creating measurements that collect partial knowledge about what occurred, like how You got traditional entanglement by removing certain possibilities. You have one q-on in the simplest embedded stories, which You track at two separate periods. You can visualize scenarios where You conclude that either your q-on form was always square or round at both times. So that all alternatives are left in play by your observations. This is the simplest case of entanglement of quantum temporal analog already explained above.

Using a somewhat more elaborate procedure, you may apply the wrinkle of complementarity to this framework. You can describe conditions that bring out the "many worlds" feature of quantum theory. Thus, at an earlier point, your q-on could be prepared in the red state, and later it will be tested to be in the blue state. As in the basic examples above, you do not reliably apply the color property to your q-on, nor does it have a certain form. Histories of this type recognize the intuition that underlies the image of quantum mechanics in many worlds in a small yet regulated and precise way. A definite state will branch into historical trajectories that are mutually inconsistent, which later come together.

Erwin Schrödinger was a founder and a pioneer of quantum theory. He was profoundly skeptical of the quantum theory's validity, emphasizing that the evolution of quantum structures inevitably leads to states that may be tested with somewhat different properties. His "Schrödinger cat" notes, famously, explain quantum uncertainty is escalating into feline mortality problems. Before calculation, as You've demonstrated in your illustrations, one does not allocate the property of existence (or death) to the cat. Inside a netherworld of probability, both or neither coexists.

Everyday language is poorly adapted for describing quantum complementarity, partially because it is not experienced in everyday practice. Practical cats deal with certain air molecules in many different forms, among other items, based on whether they are alive or dead. Still, the calculation is performed immediately in nature, and the cat goes on with its existence (or death). Yet intertwined histories define q-ons that are Schrödinger kittens in a real sense. Their complete explanation includes that You take into consideration two-contrasting property-trajectories at intermediate times.

It is delicate to monitor the experimental realization of intertwined histories since it demands that You obtain partial data regarding your q-on. Conventional quantum measurements usually collect full details at one moment, such as specifying a definite shape or a definite hue, rather than many times spanning partial data. However, without considerable technological complexity, it can be achieved. In this sense, in quantum theory, you may give definite mathematical and experimental significance to the proliferation of "many worlds" and prove its substantiality.

Chapter 7: EPR Paradox in Physics

7.1 How EPR Paradox Designates the Quantum Entanglement

EPR paradox is the thought experimentation planned in initial constructions of quantum theory to explain an intrinsic paradox. It's amongst the most well-known cases of quantum entanglement. This paradox includes 2 particles that, according to quantum mechanics, are intertwined with one another. Each particle is independently in an unknown condition under the Copenhagen interpretation of quantum mechanics before it is calculated, at which stage the state of that particle becomes definite.

The other particle's condition, therefore, becomes certain at the very same moment. This is known as a phenomenon because it appears to require interaction between the two particles with speeds higher than the speed of light. It arises a conflict with Albert Einstein's theory of relativity.

7.2 The Paradox's starting

The paradox was the major focus of heated discussion between Niels Bohr & Einstein. Einstein was never satisfied with the quantum mechanics that Bohr and his colleagues created (basically based on work originated by Einstein). Einstein created the EPR paradox along with his colleagues Boris Podolsky and Nathan Rosen to demonstrate that the idea was compatible with other existing laws of physics. There were no real means for the project to be carried out now, but it was either a logical experiment or a Gedanken experiment.

A couple of years back, the physicist David Bohm changed the example of the EPR paradox to become a little simpler. (Even to experienced physicists, the initial way the paradox was portrayed was somewhat confusing.). According to Bohformulation, an unbalanced swirl 0 element perishes in two separate particles, A & B Particle, going in contradictory directions. The sum of the two current particle spins must be equal to zero since the original particle had spin 0. If the spin of Particle A is +1/2, the spin of Particle B must be -1/2. (and vice versa).

Perhaps, it is stated by Copenhagen's interpretation of quantum mechanics; no particle has a definite condition before a measurement is generated. They all are in a superposition of potential states, with an equivalent chance of getting a positive or negative spin (in this case).

7.3 The Paradox's Sense

Two working key points can make this disturbing:

1. Quantum physics claims that particles do not have a definite quantum spin until the measurement but rather in a superposition of potential states.

2. You know for sure the benefit that You'll get from calculating the spin of Particle B as soon as You calculate the spin of Particle A.

when Particle A is measured, it appears like the calculation is responsible for setting the Particle A's quantum spin; simultaneously, particle B knows its desired spin. According to Einstein, it was a direct breach of the relativity principle.

7.4 Theory of Hidden-Variables

The second argument was never even challenged by anyone; the debate was entirely connected to the first point. Bohm and Einstein advocated an alternative method named hidden-variables theory. It implied the incompleteness of quantum mechanics. From this perspective, there had to be some quantum mechanics component that was not instantly evident but required to be applied to the theory to clarify this kind of non-local influence.

Remember, as an analogy, that you have two envelopes, one containing currency. You're informed that one of them had a $5 bill, and the other had a $10 bill. If one envelope is opened and it holds a $5 note, it will confirm that the $10 bill is in the other envelope.

The concern with this analogy is that it clearly explains that quantum mechanics operates this way. Each envelope holds a particular bill in the currency's case; it can be guessed without looking inside them.

7.5 Uncertainty in the Quantum Mechanics

In quantum mechanics, confusion is not only a lack of information but a profound lack of definitive truth. According to the Copenhagen view, before the calculation is made, the particles seem to be in a superposition of all conceivable conditions (as in the case of the dead/alive cat in the thought experiment of the Schrodinger Cat). Although most scientists may have favored a world with simpler laws, no one could work out precisely what these unknown variables are or how they could be substantially integrated into the theory.

The standard Copenhagen theory of quantum mechanics

was defended by Bohr and others, which continued to be followed by the experimental proof. The reason is that the wave's function, representing the superposition of potential quantum states, appears simultaneously at all stages. Particle A and B's spinning are not separate quantities, but they can define within the equations of the quantum entanglement by the same word. The moment the calculation is rendered on Particle A, the wave's whole function falls into a single state. There's no remote interaction taking place in this way.

7.6 Bell's Theorem

The major role acting as the last nail in the coffin of the hidden-variables theory was played by the scientist John Stewart Bell. It is regarded as Bell's Theorem. He established a set of inequalities (called Bell inequalities), representing how the spin of Particle A and Particle B would be distributed when they are not entangled. The Bell inequalities are broken in experiment after experiment, indicating that quantum entanglement does appear to occur.

Regardless of this suggestion, there's still some hidden variables theory; however, this is generally amongst incompetent scientists rather than specialists.

Chapter 8: Quantum theory: Evolution of Quantum Theory

While the theory of relativity was mostly the result of one man, Albert Einstein, with several physicists' efforts, the quantum theory was established over thirty years. The first contribution was Max Planck's 1900 description of blackbody radiation, which suggested that energies of every harmonic oscillator, like atoms of the black-body radiator, are constrained to certain quantities. Each one of which is an integral (whole number) multiple of the minimal basic value. The energy E of this simple quantum is directly proportional to the oscillator frequency v, or E=hv, where h is a constant with a value of 6.62607×10^{-34} joule-second, now named Planck's constant. In 1905, Einstein suggested that, according to the same formula, the radiation itself was also quantized, and he used the modern theory to describe the photoelectric effect. After Rutherford's (1911) observation of the nuclear atom, Bohr used the quantum theory in 1913 to describe both atomic structure and atomic spectra. It explains the relation between the electrons' energy levels and the wavelengths of light absorbed and given away.

During the 1920s, quantum mechanics, the final

mathematical formulation of quantum theory, was developed. In 1924, Louis de Broglie indicated that, as in the atomic spectra and photoelectric effect, light waves not only often exhibit particle-like properties, but particles can often manifest wave-like qualities. This hypothesis was experimentally verified in 1927 by C. L. and J. Davisson. H. Germer noticed the diffraction of an electron beam like that of a light beam. Following the recommendation of de Broglie, two separate formulations of quantum mechanics are introduced. Erwin Schrödinger's (1926) wave mechanics requires using a statistical entity, the wave function, which is linked to the possibility of locating a particle in space at a given point. According to iv's matrix mechanics, Heisenberg's given in 1925 does not discuss wave functions or related ideas. Still, it has shown that mathematically it identical to the theory of Schrödinger.

In the formulation of P. A. M. Dirac (1928), quantum mechanics was merged with relativity theory. In turn, it anticipated the presence of antiparticles. The concept of ambiguity, formulated by Heisenberg in 1927, which sets an absolute theoretical limit on the precision of such calculations, is a particularly significant revelation of quantum theory. Therefore, earlier scientists' expectation that the physical state of a system could be precisely determined and used to forecast future states had to be discarded. The theory's other advances involve quantum statistics, introduced by Einstein and S N. Bose (statistics of Bose-Einstein). It's another form. It is given by Dirac and Enrico Fermi (statistics of Fermi-Dirac), quantum electrodynamics, dealing with correlations between electromagnetic fields and charged particles. Further, it deals with generalization, quantum electronics, and the principle of quantum fields.

8.1 Five Practical Usages for "Spooky" the Quantum Mechanics

After Fifty Years of Bell's Theorem, tools that attach the properties of quantum mechanics are at work nowadays.

The quantum realm may appear to resist common sense.

Quantum Mechanics is strange. The theory that explains the functioning of small forces and particles. It notoriously left Albert Einstein so anxious that he and his colleagues stated in 1935 that it had to be unfinished; to be true, it was too "spooky."

8.2 Scientists Capture Schrödinger's Cat at the Camera

Quantum physics may be a mysterious field in which objects act in ways that do not make sense initially. The thought experiment by "Schrödinger's Cat" was intended to show some of these features, namely that molecules and atoms will live simultaneously in two distinct states before anyone looks at them. Today, this quantum paradox has been used by a community of experts to capture the intricate inner workings of molecules' insides in greater depth than it has ever been.

8.3 Seven Simple methods You'll Recognize Einstein was Correct (For Now)

Over 100 years, the general theory of relativity by Albert Einstein has survived just about any challenge that physicists have tossed at it. The popular scientist's field equations revealed in November 1915, built on the long-standing laws of Isaac Newton. It is done by redefining gravity as a covering of the framework of space and time instead of a mere force among objects.

If the masses concerned are not very huge and the speeds are reasonably tiny compared to light speed. The consequences of utilizing general relativity calculations generally appear close to what You get using Newton's calculus. Yet the idea leads to physics' revolution

Warped space-time indicates that gravity impacts light itself even more intensely than expected by Newton. It also suggests the planets travel in a somewhat altered manner through their orbits. It forecasts the presence of exotic entities such as wormholes. and monster black holes.

Einstein's gravity laws appear to fall as You relate them to the rules of quantum mechanics that reign on subatomic levels. General relativity is not flawless. In your understanding of the universe, that leaves plenty of tantalizing gaps. Scientists are stretching the boundaries even now to see exactly how deep relativity will drive us. In the meantime, there are some ways in which You see relativity in motion consistently:

8.4 Mercury's Orbit

The MESSENGER spacecraft was the first to orbit around Mercury. It took this false-color view of the small planet to display chemical, physical, and mineralogical changes on the surface.

The astronomer Urbain LeVerrier discovered a problem with

Mercury's orbit back in the 19th century. Planetary paths are not spherical; they are ellipses. It indicates that planets can be nearer or further from the sun and each other when they pass around the solar system. As planets linearly pull on each other, their closest approach points change, a phenomenon called precession.

Although even after considering the influence on all the other worlds, Mercury appeared to be moving a little faster than it was meant to do every century. At first, scientists assumed that there had to be another unknown world named Vulcan within Mercury's orbit, introducing a gravitational pull to its mix.

To demonstrate that no mystery planet was required, Einstein used the equations of general relativity. Being nearest to the earth, Mercury is more influenced by how your giant star curves the space-time fabric; Newtonian physics did not know it.

8.5 Bending Light

General relativity states that light traveling across the fabric's space-time can obey the curves of the fabric. This implies there should be light flowing around large structures bending around them. It was not clear how to observe this distortion when Einstein released his general relativity articles because the expected impact was minimal.

The British astronomer Arthur Eddington came up with an idea: to gaze at the stars near the sun's edge during the Solar Eclipse. Astronomers could see how a star's perceived location had shifted when the huge sun's gravity distorted the rays, with the glare of the sun obscured by the moon. The researchers reported from two locations: one in eastern Brazil and one in Africa.

Fair enough, after an eclipse in 1919, the Eddington squad witnessed the displacement, and media reports heralded to the nation that Einstein was right. In recent years, fresh data analyses have found that the experiment was inaccurate by contemporary standards—there are issues with the photographic plates. The quality available in 1919 was not sufficient to demonstrate the proper amount of deflection in Brazilian measurements. But later studies have shown the impact, and the study was strong enough considering the lack of modern facilities.

Nowadays, Astronomers will see the light from remote galaxies being distorted and magnified by other galaxies. It is done by utilizing strong telescopes, a phenomenon now termed gravitational lensing. This same

method is currently used to measure galaxy sizes, scan for dark matter, and check for planets circling other stars.

8.6 Black Holes

The discovery of black holes' presence shows that their structures are so large that not even light might escape their gravitational force. It is probably the most spectacular prediction of general relativity. However, the notion was not fresh. It was proposed at the Royal Society meetings in 1784 by an English physicist called John Mitchell. In 1799, French mathematician Pierre Simon Laplace came at the same notion and published a mathematically rigorous argument. Yet, no one had found something like a black hole. Moreover, observations in 1799 later appeared to demonstrate that light had to be a wave rather than a photon because gravity does not impact it in the same manner.

Introduce Einstein if gravity is simply due to curvature of space and time so that light might be influenced. Karl Schwarzschild used Einstein's equations in 1916 to illustrate that there might not only be black holes but that the resultant object was the same as Laplace's. The idea of an event horizon explains a surface from which no material

substance could escape was also developed by Schwarzschild.

While the mathematics of Schwarzschild was valid, it cost astronomers ages to detect any applicant. In the 1970s, Cygnus X-1, a potent source of X-rays, became the first entity generally recognized as a black hole. Astronomers still conclude that every galaxy has at its center a black hole, including your own. Observers carefully tracked the motions of stars around another bright X-ray while following in the center of the Milky Way, Sagittarius A; they discovered that the system acts like an incredibly large black hole.

8.7 Shooting Moon

While constructing his general relativity principle, Einstein discovered that the curve of space-time induces both the effects of gravity and the effects of motion. Besides this, the gravitational force encountered by someone standing on a huge object will be identical to the impact experienced by an individual traveling on a rocket.

It ensures that the laws of physics as determined in a laboratory will still appear the same. It is although how rapidly the laboratory travels and where it lies in space-time. Whenever you place an object in a gravitational field,

only its original location and its velocity can depend on its motion. The second statement is significant since it ensures that the tug of the gravity of the sun on earth and the moon should remain quite constant. Who knows what trouble might occur if your world and the moon "fall" at different speeds into the sun?

In the 1960s, reflectors on the moon were put up through Apollo missions & Soviet lunar inquiries. The physicists on earth shot laser rays on them to perform a variety of experimental tests. The tests include determining the gap between the moon and the earth and their comparative movement around the sun. Most of the lessons from this discovery of the lunar range were that earth and the moon fell at the same pace into the sun, as predicted by general relativity.

8.8 Dragging Space

In certain general relativity explanations, people visualize earth like a bowling ball floating on a sheet of cloth, aka space-time. The ball allows the cloth to distort into a depression. But as the Planet revolves, general relativity states that depression can twist and distort when the ball spins.

Gravity Probe B was a satellite that was set in working in 2004. It spent a year evaluating the curvature of space-time across the Planet. It discovered some proof for frame-dragging or Planet earth dragging the celestial framework with it. It helped to affirm Einstein's gravity picture.

8.9 Space-Time Ripples

Another effect of artifacts traveling across space-time is that they can also produce waves and ripples through the fabric. These gravitational waves can be potentially measurable, and they would extend space-time in an observable manner. For example, between two sets of mirrors, shine a laser beam and calculate how long it takes the beam to bounce among them. Such detectors could see a little contraction and lengthening of the beam as a space-time ripple travels across the earth, which will pop up as an interference feature.

Today, gravitational waves are considered as few basic general relativity predictions that are about to be seen, while some reports show their observation at a laboratory in the U.S. But some indirect proof remains. Pulsars are extinct stars that cram the mass of the sun into a vacuum. Some suggestions are made from observations of two pulsars

circling each other, proving that gravitational waves exist.

8.10 GPS

Global Positioning Systems are not exactly a test of relativity, but they rely on it. GPS utilizes the system of encircling satellites that ping the signals to phones & leased cars worldwide. These satellites must know the time and exact location, so they keep time measurements to high accuracy.

These satellites are orbiting 12,550 miles over your heads, where they experience less of the planet's gravitational force than humans on the earth. Einstein's principle of special relativity states that time moves differently with observers traveling at varying speeds. The satellite clocks tick a little slower than the clock on an earthbound passenger.

8.11 Physics Can Brand the Invisible Cat, Visible

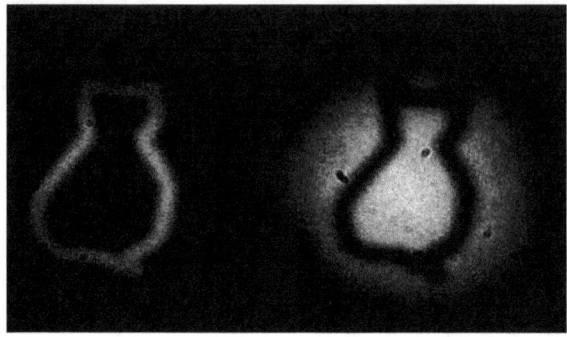

The image was produced using light particles (photons)—that have been quantum entangled. It means that these pairs of particles share properties.

In this experiment, researchers broke a green laser beam into pieces and changed the wavelengths such that one beam was yellow and the other was red. Laser light contains only one color of light with both troughs and peaks of the light waves lined up.) Piece of the frame entangled particles between those two rays. Next, red photons were sent to the cat stencil, and the camera received the yellow photons.

The cat cut-outs have been constructed on a film opaque to yellow light. As the light rays, you're intertwined, the yellow photons "showed" the cat's picture to the camera while the red photons might "see" the stencil simply.

The researchers selected a cat outline to pay homage to an experiment suggested by Austrian physicist Erwin Schrödinger. He was the scientist who came with the word entanglement. His theory, the "Schrödinger's cat" paradox, addresses the strange implications of the quantum entanglement: In his illustration, a cat is in a tank with a flask of radioactive gas that has a 50/50 percent risk of leakage.

The cat would be expired after some time. Or it will live. But

before You look, you have no idea if the cat is dead or alive. Therefore, the cat lives in a superposition between both dead and living. Your discovery causes the options to fall into one reality.

8.12 Ultra-Precise Clocks

Accurate timekeeping is for more than just the morning reminder. Clocks coordinate your technical environment, holding stuff like financial exchanges and GPS networks in line. Normal clocks utilize the usual alternations of physical artifacts like quartz or pendulums crystals to create their 'ticks' & 'tocks.' Nowadays, the most reliable clocks worldwide are atomic clocks, which were able to use concepts of quantum theory to calculate time. They track the exact radiation frequency required to allow electrons to move between energy levels. The quantum-logic clock at the U.S.

Global Institute of Standards and Technology (NIST) in Colorado just loses or wins a second per 3.7 billion years. NIST strontium clock, released earlier this year, would be reliable for 5 billion years—longer than the Planet's actual age. These incredibly sensitive atomic clocks assist with GPS navigation, communications, and observing.

8.13 Uncrackable Codes

Traditional cryptography operates with keys: A sender uses one key to decrypt the content, and a receiver uses another to decode the document. However, it's impossible to eliminate the danger of an eavesdropper, and keys may be compromised. This can be resolved using theoretically impenetrable quantum key distribution (QKD). In QKD, knowledge about the key is transmitted by photons that have been randomly polarized. This limits the photon such that it vibrates in just one plane—for, e.g., up, and down, or left to right. The receiver will use polarized filters to decode the key and then use a preferred algorithm to successfully encrypt a document.

The coded data always gets transmitted through standard communication lines,

but no one will decipher the message until they have the specific quantum key. That's complicated since quantum

laws say that "reading" the polarized photons will still alter their states, and any effort at eavesdropping will warn the participants of a security violation.

8.14 Super- Computers

A standard computer encrypts data as bits or a series of binary digits; Quantum computers supercharge computing capacity when they utilize quantum bits or qubits and reside in a superposition of states—until they are counted, qubits maybe both "1" and "0" at the same time.

This sector is in progress, but there have been moves in the appropriate direction. In 2011, D-Wave Devices announced the D-Wave One, a 128-qubit processor, preceded a year later by the 512-qubit D-Wave Two. The corporation claims they are the world's first publicly available quantum computers. However, this argument has been met with doubt, in part because it's still uncertain if D-Wave's qubits are intertwined. Studies published in May found signs of entanglement but only in a limited subset of the computer's qubits. There's still confusion about whether the chips show any accurate quantum speed. NASA and

Google have partnered up to form the Quantum Artificial Intelligence Lab focused on a D-Wave Two. And scientists

at the University of Bristol last year linked up one of their conventional quantum chips to the Internet so anybody with a web browser would practice quantum coding.

8.15 Improved Microscopes

In February, a group of investigators at Japan's Hokkaido University created the world's first entanglement-enhanced

microscope. They utilize a technique called differential interference contrast microscopy. This microscope shoots two beams of photons at a material and tests the mirrored beam's interference pattern. The pattern varies based on whether they strike a smooth or irregular surface. Using entangled photons significantly enhances the amount of knowledge the microscope may collect since calculating one entangled photon gives data about its companion.

The Hokkaido team has managed to imagine an etched "Q" that held only 17 nanometers above the surface with unparalleled sharpness. Similar methods may be used to boost the resolution of astronomy instruments called interferometers, which superimpose various waves of light to further evaluate their properties. Interferometers are

used in the quest for extrasolar worlds. It is used to probe neighboring stars and check for ripples in spacetime, known as gravitational waves.

8.16 Biological Compasses

Humans aren't the only people making use of quantum mechanics. One leading hypothesis indicates that birds like the European robin use some mysterious behavior to stay on track as they migrate.

The approach requires a light-sensitive protein named cryptochrome, which can include entangled electrons. When photons penetrate the eye, they reach the cryptochrome molecules. They may provide enough energy to tear them apart, creating two reactive molecules, or radicals, with unpaired but still intertwined electrons. The magnetic field around the bird determines how long these cryptochrome radicals survive. Cells in the bird's retina are believed to be very receptive to the existence of the intertwined radicals, enabling the animals to essentially 'see' a magnetic diagram dependent on the molecules.

This mechanism isn't completely known; however, there is another issue: Birds' magnetic responsiveness may be attributed to tiny crystals of magnetic minerals found in their

beaks. Studies show that if entanglement is at play, the delicate condition may persist even for longer in a bird's

eye than in all the strongest artificial systems. The magnetic compass may even be useful to some lizards, insects, crustaceans, and even mammals. For instance, a type of cryptochrome used for magnetic navigation in flies has also been detected in the human eye: although it's uncertain whether it is or once was useful for a similar reason.

Chapter 9: Where, in the Future, Is the Quantum Computing?

What do rechargeable batteries, prescription medications, and solar cells have familiar with? They share the ability to gain great benefits in design from the simulation of their behavior at the stage of quantum mechanics. The problem is, modeling the quantum dynamics of these devices may be incredibly complex for even the biggest supercomputers. The simulation involves keeping track of and conducting calculations on various variables that expand with the number of electrons in every molecule.

Full quantum-mechanical implementation will take even a supercomputer's thousands of years to complete for certain systems. In the early 1980s, this challenge of modeling quantum systems motivated physicist Richard Feynman to introduce a device that works quantum mechanically in a significant way.

9.1 What is Next in the Computing? Exploring Possibilities

McMahon's analysis centers around utilizing physical devices to do computing in better format than the modern means, employing traditional machines. Along with investigating the potential of quantum computers, the McMahon lab often explores alternate classical computing systems. It includes photonic neural networks, in which computations are conducted utilizing light instead of electrons. McMahon and his collaborators see their study as investigating what could come next. The computing industry is near reaching the boundary for the scaling downsize of the transistors in the silicon chips, which is the main primary driver of change in computer efficiency.

Instead of seeking to find methods to push existing computer processors' capability, the researchers are addressing the issue from a more basic, applied physics. "Let's stand back and ask, 'when You had to do this all again, what would You do?'" McMahon replied. "What is the most effective way to generate a processor?"

9.2 Quantum Way

In quantum computing systems, numerous technological candidates create the basic units of information, known as quantum bits or qubits. The McMahon lab is presently focusing on two major things, including photonics and superconducting circuits.

"In all of these technologies, you have a lot of ideas You'd like to carry out," McMahon states. "In your photonic approach, the aim is to build a prototype quantum computer that is constrained in what it can perform but will still be a device that a classical computer could not easily start. In the near term, it will tell you just how feasible your photonic architecture is to understand. And in a more remote future, you will theoretically create devices that are a fundamental quantum computer that is capable of operating any algorithm."

"How do you acquire photonic qubits to communicate with each other? It's possible to make this work, but they don't want to."

To build quantum computers, McMahon and his collaborators would need to address the inherent limitations of the technology they are using. Photonic qubits, for example, usually don't communicate with each other. "That is a huge issue for quantum computation," McMahon notes. "How do you make photonic qubits to talk with each other?" It's possible to force them to do so; however, they don't naturally want to, especially in certain situations.

9.3 Superconducting Circuits

On the other hand, Superconducting circuits present their issues. It may seem an easy skill to use as they look extraordinarily like classical electronic circuits. A major hook is that they operate at a temperature close to absolute zero.

"If they don't run at very low temperature, they don't act as quantum-mechanical systems," McMahon states. "In other words, the impact of quantum-mechanical gets washed out." Keeping the circuits sufficiently cold while increasing their number and the complexity of computations they perform is a tall order.

9.4 What Can You Do with a Quantum Computer?

McMahon also needs to investigate the possible applications for a quantum machine, including and beyond Feynman's initial proposal that they utilized for simulating the quantum systems. To that terminal, he also is working with Paul Gainshare & Thomas Hartman, Physics. "You're hoping to find anything useful You can do with a near-term quantum computer that would respond to questions in quantum gravity, or high-energy physics more specifically, that couldn't be resolved otherwise," McMahon says. "For instance, can You visualize a model of a black hole on a quantum computer? Will it be useful? You don't know if You'll discover it, so it's really fun to try."

9.5 Using Light versus Electrons for Processors

One of the main tasks a processor performs for machine learning and specifically for neural networks—is matrix-vector multiplication. All these multiplications model knowledge transfer through levels of the artificial neurons. Traditional computers that operate current neural network algorithms utilize a large portion of computing power to conduct matrix calculations. McMahon addresses matrix-vector multiplication by constructing photonic processors that conduct the multiplications and additions using light instead of electrons. It is a continuation of his exploration of physical structures that are naturally suitable for computation.

"If you thought of a light beam, you could interpret different parts of the beam that encodes different elements of a vector, or equally in the case of neural networks, various neuron values," McMahon says. "As a result of how light propagation works, if You send light via a designed medium or optical apparatus that allows various portions of an incoming beam disperse in distinct ways by different amounts, you can describe what occurs as a matrix-vector multiplication. So only by shining the light on the carefully built optical instrument, you may achieve the desired matrix-vector multiplication."

Conclusion

This book is a guide for beginners to untie the basic ambiguities of quantum entanglement. It is a wide-ranging book to aid people in realizing it better. Quantum physics plays an essential part in your lives. It is very significant for everyone to have at least the basic information on this subject. Mostly people struggle with it as there are hardly any books on these topics compatible with people's demands. Especially those who are just starting their career as physicists and need a guide to understand the concepts. The objective of the e-Book is simple which is to assist individuals in having the better considerate of quantum physics simplest of way. It provides learning about: Relation between waves & particles, So, Why Max Planck considered the father of the Quantum Physics, quantum physics Laws, theory of the Quantum field, relative theory of Einstein, Hydrogen atom Importance Basics on the angular momentum at the quantum. Firstly, you all need to understand that all the things in this whole universe are composed of particles and waves. This is known as the double nature of substance because they both exist at the same time. It seems like a miracle and simply hard to accept. These results have been obtained from many scientific experiments. Secondly, you must understand and accept that it is impossible to predict an experiment's exact result on a quantum system when it comes to quantum physics. There is only probability but no certainty that leads us to conclude that quantum physics is probabilistic. Finally, you must understand the small size of the quantum entanglement is very small. It explains that quantum physics is worked when objects are extremely small. Its reason is quantum effects that play a role in the procedure get smaller with the increase in the object's size. Therefore, quantum behaviors are difficult to discover.

***I'm glad** you made it this far! It's very important to me to know if this book has done its job and if you feel a little more cultured now, so please, leave me a short review.*

Thanks for spending your time

GIFT - download the e-book version for FREE:

www.bit.ly/qpebook

www.ingramcontent.com/pod-product-compliance
Lightning Source LLC
LaVergne TN
LVHW011709060526
838200LV00051B/2819